HOLT®

ChemFile® C

Inquiry Experiments
Teacher Guide

HOLT, RINEHART AND WINSTON

A Harcourt Education Company

Orlando • **Austin** • New York • San Diego • Toronto • London

ISBN 0-03-036803-0

2 3 4 5 6 7 8 9   170   07 06 05

Contents

Ordering Lab Materials with the
One-Stop Planner® CD-ROM

Your class and prep time are valuable. Now, it's easier and faster than ever to organize and obtain the materials that you need for all of the labs in the *Modern Chemistry* lab program. Using the Lab Materials QuickList software found on the *One-Stop Planner® CD-ROM*, you can do the following:

• View all of the materials that you need for any (or all) labs.
The Lab Materials QuickList software allows you to easily see all of the materials needed for any lab in the *Modern Chemistry* lab program. Use your materials list to order all of your materials at once. Or use the list to determine what items you need to resupply or supplement your stockroom so that you'll be prepared to do any lab.

• Create a customized materials list.
No two teachers teach exactly alike. The Lab Materials QuickList software allows you to select labs by type of lab or by type of material used. You can create a materials list that summarizes the needs of whichever labs from *Modern Chemistry* you choose.

• Let the software handle the details.
You can customize your list based on the number of students and the number of lab groups. A powerful software engine that has been programmed to distinguish between consumable and nonconsumable materials will "do the math." Whether you're examining all of the labs for a whole year or just the labs that you're planning for next week, the software does the hard work of totaling and tallying, telling you what you need and *exactly* how much you'll need for the labs that you've selected.

• Print your list.
By printing out materials lists that you created by using the Lab Materials QuickList software, you can have a copy of any materials list right at your fingertips for easy reference at any time.

• Order your materials easily.
After you've created your materials list by using the Lab Materials QuickList software, you can use it to order from Sargent-Welch or to prepare a purchase order to be sent directly to another scientific materials supplier.

Visit go.hrw.com to learn more about the Lab Materials QuickList software.

Teacher's Laboratory Safety Information

HELPING STUDENTS RECOGNIZE THE IMPORTANCE OF SAFETY

One method that can help students appreciate the importance of precautions is to use a safety contract that students read and sign, indicating they have read, understand, and will respect the necessary safety procedures, as well as any other written or verbal instructions that will be given in class. You can find a copy of a safety contract on the *Modern Chemistry One Stop Planner CD-ROM.* You can use this form as a model or make your own safety contract for your students with language specific to your lab situation. When making your own contract, you could include points such as the following:

- Make sure that students agree to wear personal protective equipment (goggles and lab aprons) at all times. Safety information regarding the use of contact lenses is continually changing. Check your state and local regulations on this subject. Students should agree to read all lab exercises before they come to class. They should agree to follow all directions and safety precautions and to use only materials and equipment that you provide.
- Students should agree to remain alert and cautious at all times in the lab. They should never leave experiments unattended.
- Students should not wear dangling jewelry or bulky clothing.
- Students should bring only lab manuals and lab notebooks into the lab. Backpacks, textbooks for other subjects, and other items should be stored elsewhere.
- Students should agree to never eat, drink, or smoke in any science laboratory. Food should **never** be brought into the laboratory
- Students should **never** taste or touch chemicals.
- Students should keep themselves and other objects away from Bunsen burner flames. Students should be responsible for checking, before they leave, that gas valves and hot plates are off.
- Students should know the proper fire drill procedures and the locations of fire exits.
- Students should always clean all apparatus and work areas.
- Students should wash their hands thoroughly with soap and water before leaving the lab room.
- Students should know the locations and operation of all safety equipment in the laboratory.
- Students should report all accidents or close calls to you immediately, no matter how minor.
- Students should **never** work alone in the laboratory, and they should never work unless you are present.

DISPOSAL OF CHEMICALS

Only a relatively small percentage of waste chemicals are classified as hazardous by EPA regulations. The EPA regulations are derived from two acts (as amended) passed by the Congress of the United States: RCRA (Resource Conservation and Recovery Act) and CERCLA (Comprehensive Environmental Response, Compensation, and Liability Act).

In addition, some states have enacted legislation governing the disposal of hazardous wastes that differs to some extent from the federal legislation. The disposal procedures described in this book have been designed to comply with the federal legislation as described in the EPA regulations.

In most cases the disposal procedures indicated in the teacher's edition will probably comply with your state's disposal requirements. However, to be sure of this, check with your state's environmental agency. If a particular disposal procedure does not comply with your state requirements, ask that agency to assist you in devising a procedure that is in compliance.

The following general practices are recommended in addition to the specific instructions given in the Lab Notes.

- Except when otherwise specified in the disposal procedures, neutralize acidic and basic wastes with 1.0 M potassium hydroxide, KOH, or 1.0 M sulfuric acid, H_2SO_4, added slowly and stirred.

- In dealing with a waste disposal contractor, prepare a complete list of the chemicals you want to dispose of. Classify each chemical on your disposal list as hazardous or nonhazardous waste. Check with your local environmental agency office for the details of such classification.

- Unlabeled bottles are a special problem. They must be identified to the extent that they can be classified as hazardous or nonhazardous waste. Some landfills will analyze a mystery bottle for a fee if it is shipped to the landfill in a separate package, is labeled as a sample, and includes instructions to analyze the contents sufficiently to allow proper disposal.

ELECTRIC SAFETY

Although none of the labs in this manual require electrical equipment, several include options for the use of microcomputer-based laboratory equipment, pH meters, or other equipment. The following safety precautions to avoid electrical shocks must be observed any time electrical equipment is present in the lab:

- Each electrical socket in the laboratory must be a three-hole socket and must be protected with a GFI (ground-fault interrupter) circuit.

- Check the polarity of all circuits with a polarity tester from an electronics supply store before you use them. Repair any incorrectly wired sockets.

- Use only electrical equipment that has a three-wire cord and three-prong plug.

- Be sure all electrical equipment is turned off before it is plugged into a socket. Turn off electrical equipment before you unplug it.

- Wiring hookups should be made or altered only when apparatus is disconnected from the power source and the power switch is turned off.

- Do not let electrical cords dangle from work stations; dangling cords are a shock hazard, and students can trip over them.
- Do not use electrical equipment with frayed or twisted cords.
- The area under and around electrical equipment should be dry; cords should not lie in puddles of spilled liquid.
- Your hands should be dry when you use electrical equipment.
- Do not use electrical equipment powered by 110–115 V alternating current for conductivity demonstrations or for any other use in which bare wires are exposed, even if the current is connected to a lower-voltage AC or DC connection.

   Use dry cells or nicad rechargeable batteries as direct-current sources. Do not use automobile storage batteries or AC-to-DC converters; these two sources of DC current can present serious shock hazards.

Prepared by Jay A. Young, Consultant, Chemical Health and Safety, Silver Spring, Maryland

Incompatible Chemicals

Consider the following list when organizing and storing chemicals. Note that some chemicals on this list should no longer be in your lab, due to their potential risks. Consult your state and local guidelines for more specific information on chemical hazard.

Chemical	Should not come in contact with
acetic acid	chromic acid, nitric acid, perchloric acid, ethylene glycol, hydroxyl compounds, peroxides, or permanganates
acetone	concentrated sulfuric acid or nitric acid mixtures
acetylene	silver, mercury, or their compounds; bromine, chlorine, fluorine, or copper tubing
alkali metals, powdered aluminum or magnesium	water, carbon dioxide, carbon tetrachloride, or the halogens
ammonia (anhydrous)	mercury, hydrogen fluoride, or calcium hypochlorite
ammonium nitrate (strong oxidizer)	strong acids, metal powders, chlorates, sulfur, flammable liquids, or finely divided organic materials
aniline	nitric acid or hydrogen peroxide
bromine	ammonia, acetylene, butane, hydrogen, sodium carbide, turpentine, or finely divided metals
carbon (activated)	calcium hypochlorite or any oxidizing agent
chlorates	ammonium salts, strong acids, powdered metals, sulfur, or finely divided organic materials
chromic acid	glacial acetic acid, camphor, glycerin, naphthalene, turpentine, low-molar-mass alcohols, or flammable liquids
chlorine	same as bromine
copper	acetylene or hydrogen peroxide
flammable liquids	ammonium nitrate, chromic acid, hydrogen peroxide, sodium peroxide, nitric acid, or any of the halogens
hydrocarbons (butane, propane, gasoline, turpentine)	fluorine, chlorine, bromine, chromic acid, or sodium peroxide
hydrofluoric acid	ammonia
hydrogen peroxide	most metals or their salts, flammable liquids, or other combustible materials
hydrogen sulfide	nitric acid or certain other oxidizing gases
iodine	acetylene or ammonia
nitric acid	glacial acetic acid, chromic or hydrocyanic acids, hydrogen sulfide, flammable liquids, or flammable gases that are easily nitrated

Chemical	Should not come in contact with
oxygen	oils, grease, hydrogen, or flammable substances
perchloric acid	acetic anhydride, bismuth or its alloys, alcohols, paper, wood, or other organic materials
phosphorus pentoxide	water
potassium permanganate	glycerin, ethylene glycol, or sulfuric acid
silver	acetylene, ammonium compounds, oxalic acid, or tartaric acid
sodium peroxide	glacial acetic acid, acetic anhydride, methanol, carbon disulfide, glycerin, benzaldehyde, or water
sulfuric acid	chlorates, perchlorates, permanganates, or water

Introduction to the Lab Program

Structure of the Experiments

INTRODUCTION

The opening paragraphs set the theme for the experiment and summarize its major concepts.

OBJECTIVES

Objectives highlight the key concepts to be learned in the experiment and emphasize the science process skills and techniques of scientific inquiry.

MATERIALS

These lists enable you to organize all apparatus and materials needed to perform the experiment. Knowing the concentrations of solutions is vital. You often need this information to perform calculations and to answer the questions at the end of the experiment.

SAFETY

Safety cautions are placed at the beginning of the experiment to alert you to procedures that may require special care. Before you begin, you should review the safety issues that apply to the experiment.

PROCEDURE

By following the procedures of an experiment, you perform concrete laboratory operations that duplicate the fact-gathering techniques used by professional chemists. You learn skills in the laboratory. The procedures tell you how and where to record observations and data.

DATA AND CALCULATIONS TABLES

The data that you collect during each experiment should be recorded in the labeled Data Tables provided. The entries you make in a Calculations Table emphasize the mathematical, physical, and chemical relationships that exist among the accumulated data. Both types of tables should help you to think logically and to formulate your conclusions about what occurs during the experiment.

CALCULATIONS

Space is provided for all computations based on the data you gather.

QUESTIONS

Based on the data and calculations, you should be able to develop plausible explanations for the phenomena you observe during the experiment. Specific questions require you to draw on the concepts you learn.

GENERAL CONCLUSIONS

This section asks broader questions that bring together the results and conclusions of the experiment and relate them to other situations.

Safety in the Chemistry Laboratory

CHEMICALS ARE NOT TOYS.

Any chemical can be dangerous if it is misused. Always follow the instructions for the experiment. Pay close attention to the safety notes. Do not do anything differently unless told to do so by your teacher.

Chemicals, even water, can cause harm. The trick is to know how to use chemicals correctly so that they will not cause harm. You can do this by following the rules on these pages, paying attention to your teacher's directions, and following the cautions on chemical labels and experiments.

These safety rules always apply in the lab.

1. **Always wear a lab apron and safety goggles.**
   Even if you aren't working on an experiment, laboratories contain chemicals that can damage your clothing, so wear your apron and keep the strings of the apron tied. Because chemicals can cause eye damage, even blindness, you must wear safety goggles. If your safety goggles are uncomfortable or get clouded up, ask your teacher for help. Try lengthening the strap a bit, washing the goggles with soap and warm water, or using an antifog spray.

2. **Generally, no contact lenses are allowed in the lab.**
   Even while wearing safety goggles, you can get chemicals between contact lenses and your eyes, and they can cause irreparable eye damage. If your doctor requires that you wear contact lenses instead of glasses, then you may need to wear special safety goggles in the lab. Ask your doctor or your teacher about them.

3. **Never work alone in the laboratory.**
   You should always do lab work under the supervision of your teacher.

4. **Wear the right clothing for lab work.**
   Necklaces, neckties, dangling jewelry, long hair, and loose clothing can cause you to knock things over or catch items on fire. Tuck in a necktie or take it off. Do not wear a necklace or other dangling jewelry, including hanging earrings. It isn't necessary, but it might be a good idea to remove your wristwatch so that it is not damaged by a chemical splash.

   Pull back long hair, and tie it in place. Nylon and polyester fabrics burn and melt more readily than cotton, so wear cotton clothing if you can. It's best to wear fitted garments, but if your clothing is loose or baggy, tuck it in or tie it back so that it does not get in the way or catch on fire.

   Wear shoes that will protect your feet from chemical spills—no open-toed shoes, sandals, or shoes made of woven leather straps. Shoes made of solid leather or a polymer are much better than shoes made of cloth. Also, wear pants, not shorts or skirts.

5. **Only books and notebooks needed for the experiment should be in the lab.**
   Do not bring other textbooks, purses, bookbags, backpacks, or other items into the lab; keep these things in your desk or locker.

6. **Read the entire experiment before entering the lab.**
Memorize the safety precautions. Be familiar with the instructions for the experiment. Only materials and equipment authorized by your teacher should be used. When you do the lab work, follow the instructions and the safety precautions described in the directions for the experiment.

7. **Read chemical labels.**
Follow the instructions and safety precautions stated on the labels. Know the location of Material Safety Data Sheets for chemicals.

8. **Walk carefully in the lab.**
Sometimes you will carry chemicals from the supply station to your lab station. Avoid bumping other students and spilling the chemicals. Stay at your lab station at other times.

9. **Food, beverages, chewing gum, cosmetics, and tobacco are *never* allowed in the lab.**
You already know this.

10. **Never taste chemicals or touch them with your bare hands.**
Also, keep your hands away from your face and mouth while working, even if you are wearing gloves.

11. **Use a sparker to light a Bunsen burner.**
Do not use matches. Be sure that all gas valves are turned off and that all hot plates are turned off and unplugged before you leave the lab.

12. **Be careful with hot plates, Bunsen burners, and other heat sources.**
Keep your body and clothing away from flames. Do not touch a hot plate just after it has been turned off. It is probably hotter than you think. Use tongs to heat glassware, crucibles, and other things and to remove them from a hot plate, a drying oven, or the flame of a Bunsen burner.

13. **Do not use electrical equipment with frayed or twisted cords or wires.**

14. **Be sure your hands are dry before you use electrical equipment.**
Before plugging an electrical cord into a socket, be sure the electrical equipment is turned off. When you are finished with it, turn it off. Before you leave the lab, unplug it, but be sure to turn it off first.

15. **Do not let electrical cords dangle from work stations; dangling cords can cause tripping or electric shocks.**
The area under and around electrical equipment should be dry; cords should not lie in puddles of spilled liquid.

16. **Know fire drill procedures and the locations of exits.**

17. **Know the locations and operation of safety showers and eyewash stations.**

18. **If your clothes catch on fire, *walk* to the safety shower, stand under it, and turn it on.**

19. **If you get a chemical in your eyes, walk immediately to the eyewash station, turn it on, and lower your head so that your eyes are in the running water.**
Hold your eyelids open with your thumbs and fingers, and roll your eyeballs around. You have to flush your eyes continuously for at least 15 min. Call your teacher while you are doing this.

20. **If you have a spill on the floor or lab bench, don't try to clean it up by yourself.**
First, ask your teacher if it is OK for you to do the cleanup; if it is not, your teacher will know how the spill should be cleaned up safely.

21. **If you spill a chemical on your skin, wash it off under the sink faucet, and call your teacher.**
If you spill a solid chemical on your clothing, brush it off carefully so that you do not scatter it, and call your teacher. If you get a liquid chemical on your clothing, wash it off right away if you can get it under the sink faucet, and call your teacher. If the spill is on clothing that will not fit under the sink faucet, use the safety shower. Remove the affected clothing while under the shower, and call your teacher. (It may be temporarily embarrassing to remove your clothing in front of your class, but failing to flush that chemical off your skin could cause permanent damage.)

22. **The best way to prevent an accident is to stop it before it happens.**
If you have a close call, tell your teacher so that you and your teacher can find a way to prevent it from happening again. Otherwise, the next time, it could be a harmful accident instead of just a close call.

23. **All accidents should be reported to your teacher, no matter how minor.**
Also, if you get a headache, feel sick to your stomach, or feel dizzy, tell your teacher immediately.

24. **For all chemicals, take only what you need.**
On the other hand, if you do happen to take too much and have some left over, **do not** put it back in the bottle. If somebody accidentally puts a chemical into the wrong bottle, the next person to use it will have a contaminated sample. Ask your teacher what to do with any leftover chemicals.

25. *Never* **take any chemicals out of the lab.**
You should already know this rule.

26. **Horseplay and fooling around in the lab are very dangerous.**
Never be a clown in the laboratory.

27. **Keep your work area clean and tidy.**
After your work is done, clean your work area and all equipment.

28. **Always wash your hands with soap and water before you leave the lab.**

29. **Whether or not the lab instructions remind you, *all* of these rules *apply all of the time.***

QUIZ

Determine which safety rules apply to the following.

- Tie back long hair, and confine loose clothing. (Rule ? applies.)
- Never reach across an open flame. (Rule ? applies.)
- Use proper procedures when lighting Bunsen burners. Turn off hot plates and Bunsen burners that are not in use. (Rule ? applies.)
- Be familiar with the procedures and know the safety precautions before you begin. (Rule ? applies.)
- Use tongs when heating containers. Never hold or touch containers with your hands while heating them. Always allow heated materials to cool before handling them. (Rule ? applies.)
- Turn off gas valves that are not in use. (Rule ? applies.)

SAFETY SYMBOLS

To highlight specific types of precautions, the following symbols are used in the experiments. Remember that no matter what safety symbols and instructions appear in each experiment, all of the 29 safety rules described previously should be followed at all times.

EYE AND CLOTHING PROTECTION

- Wear safety goggles in the laboratory at all times. Know how to use the eyewash station.

- Wear laboratory aprons in the laboratory. Keep the apron strings tied so that they do not dangle.

CHEMICAL SAFETY

- Never taste, eat, or swallow any chemicals in the laboratory. Do not eat or drink any food from laboratory containers. Beakers are not cups, and evaporating dishes are not bowls.
- Never return unused chemicals to their original containers.
- Some chemicals are harmful to the environment. You can help protect the environment by following the instructions for proper disposal.
- It helps to label the beakers and test tubes containing chemicals.
- Never transfer substances by sucking on a pipet or straw; use a suction bulb.
- Never place glassware, containers of chemicals, or anything else near the edges of a lab bench or table.

CAUSTIC SUBSTANCES

- If a chemical gets on your skin or clothing or in your eyes, rinse it immediately, and alert your teacher.
- If a chemical is spilled on the floor or lab bench, tell your teacher, but do not clean it up yourself unless your teacher says it is OK to do so.

HEATING SAFETY

- When heating a chemical in a test tube, always point the open end of the test tube away from yourself and other people.

EXPLOSION PRECAUTION

- Use flammable liquids in small amounts only.
- When working with flammable liquids, be sure that no one else in the lab is using a lit Bunsen burner or plans to use one. Make sure there are no other heat sources present.

HAND SAFETY

- Always wear gloves or use cloths to protect your hands when cutting, fire polishing, or bending hot glass tubing. Keep cloths clear of any flames.
- Never force glass tubing into rubber tubing, rubber stoppers, or corks. To protect your hands, wear heavy leather gloves or wrap toweling around the glass and the tubing, stopper, or cork, and gently push the glass tubing into the rubber or cork.
- Use tongs when heating test tubes. Never hold a test tube in your hand to heat it.
- Always allow hot glassware to cool before you handle it.

GLASSWARE SAFETY

- Check the condition of glassware before and after using it. Inform your teacher of any broken, chipped, or cracked glassware because it should not be used.
- Do not pick up broken glass with your bare hands. Place broken glass in a specially designated disposal container.

GAS PRECAUTION

- Do not inhale fumes directly. When instructed to smell a substance, waft it toward you. That is, use your hand to wave the fumes toward your nose. Inhale gently.

RADIATION PRECAUTION

- Always wear gloves when handling a radioactive source.
- Always wear safety goggles when performing experiments with radioactive materials.
- Always wash your hands and arms thoroughly after working with radioactive materials.

HYGIENIC CARE

- Keep your hands away from your face and mouth.
- Always wash your hands before leaving the laboratory.

Any time you see any of the safety symbols, you should remember that all 29 of the numbered laboratory rules always apply.

Labeling of Chemicals

In any science laboratory the labeling of chemical containers, reagent bottles, and equipment is essential for safe operations. Proper labeling can lower the potential for accidents that occur as a result of misuse. Read labels and equipment instructions several times before you use chemicals or equipment. Be sure that you are using the correct items, that you know how to use them, and that you are aware of any hazards or precautions associated with their use.

All chemical containers and reagent bottles should be labeled prominently and accurately with labeling materials that are not affected by chemicals.

Chemical labels should contain the following information:

1. **Name of the chemical and its chemical formula**

2. **Statement of possible hazards** This is indicated by the use of an appropriate signal word, such as *DANGER*, *WARNING*, or *CAUTION*. This signal word usually is accompanied by a word that indicates the type of hazard present, such as *POISON*, *CAUSES BURNS*, *EXPLOSIVE*, or *FLAMMABLE*. Note that this labeling should not take the place of reading the appropriate Material Safety Data Sheet for a chemical.

3. **Precautionary measures** Precautionary measures describe how users can avoid injury from the hazards listed on the label. Examples include: "use only with adequate ventilation" and "do not get in eyes or on skin or clothing."

4. **Instructions in case of contact or exposure** If accidental contact or exposure does occur, immediate first-aid measures can minimize injury. For example, the label on a bottle of acid should include this instruction: "In case of contact, flush with large amounts of water; for eyes, rinse freely with water for 15 min and get medical attention immediately."

5. **The date of preparation and the name of the person who prepared the chemical** This information is important for maintaining a safe chemical inventory.

Suggested Labeling Scheme	
Name of contents	hydrochloric acid
Chemical formula and concentration or physical state	6 M HCl
Statements of possible hazards and precautionary and measures	WARNING! CAUSTIC and CORROSIVE—CAUSES BURNS Avoid contact with skin and eyes Avoid breathing vapors.
Hazard Instructions for contact or overexposure	IN CASE OF CONTACT: Immediately flush skin or eyes with large amounts of water for at least 15 min; for eyes, get medical attention immediately!
Date prepared or obtained Manufacturer (commercially obtained) or preparer (locally made)	May 8, 2005 Prepared by Betsy Byron, Faribault High School, Faribault, Minnesota

Laboratory Techniques

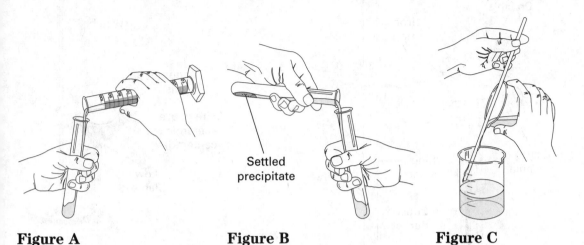

Figure A   **Figure B**   **Figure C**

Settled precipitate

DECANTING AND TRANSFERRING LIQUIDS

1. The safest way to transfer a liquid from a graduated cylinder to a test tube is shown in **Figure A.** Transfer the liquid at arm's length with your elbows slightly bent. This position enables you to see what you are doing and still maintain steady control.

2. Sometimes liquids contain particles of insoluble solids that sink to the bottom of a test tube or beaker. Use one of the methods given below to separate a supernatant (the clear fluid) from insoluble solids.

    a. Figure B shows the proper method of decanting a supernatant liquid in a test tube.

    b. Figure C shows the proper method of decanting a supernatant liquid in a beaker by using a stirring rod. The rod should touch the wall of the receiving container. Hold the stirring rod against the lip of the beaker containing the supernatant liquid. As you pour, the liquid will run down the rod and fall into the beaker resting below. Using this method will prevent the liquid from running down the side of the beaker you are pouring from.

HEATING SUBSTANCES AND EVAPORATING SOLUTIONS

1. Use care in selecting glassware for high-temperature heating. The glassware should be heat resistant.

2. When using a gas flame to heat glassware, use a ceramic-centered wire gauze to protect glassware from direct contact with the flame. Wire gauzes can withstand extremely high temperatures and will help prevent glassware from breaking. **Figure D** shows the proper setup for evaporating a solution over a water bath.

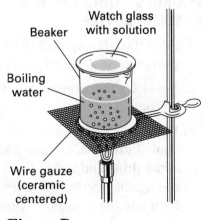

Watch glass with solution

Beaker

Boiling water

Wire gauze (ceramic centered)

Figure D

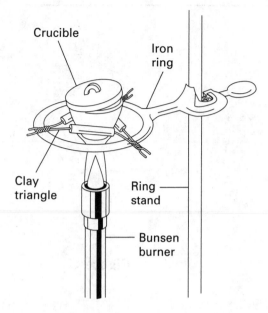

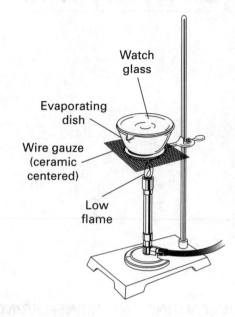

Figure E   **Figure F**

3. In some experiments you are required to heat a substance to high temperatures in a porcelain crucible. **Figure E** shows the proper apparatus setup used to accomplish this task.

4. **Figure F** shows the proper setup for evaporating a solution in a porcelain evaporating dish with a watch glass cover that prevents spattering.

5. Glassware, porcelain, and iron rings that have been heated may look cool after they are removed from a heat source, but they can burn your skin even after several minutes of cooling. Use tongs, test tube holders, or heat-resistant mitts and pads whenever you handle this apparatus.

6. You can test the temperature of questionable beakers, ring stands, wire gauzes, or other pieces of apparatus that have been heated, by holding the back of your hand close to their surfaces before grasping them. You will be able to feel any heat generated from the hot surfaces. **Do not touch the apparatus until it is cool.**

POURING LIQUID FROM A REAGENT BOTTLE

1. Read the label at least three times before using the contents of a reagent bottle.

2. Never lay the stopper of a reagent bottle on the lab table.

3. When pouring a caustic or corrosive liquid into a beaker, use a stirring rod to avoid drips and spills. Hold the stirring rod against the lip of the reagent bottle. Estimate the amount of liquid you need, and pour this amount along the rod into the beaker. See **Figure G.**

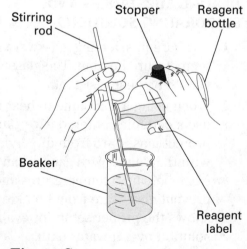

Figure G

4. Take extra precautions when handling a bottle of acid or strong base. Remember the following important rules: Never add water to any concentrated acid, particularly sulfuric acid, because the mixture can splash and will generate a lot of heat. To dilute any acid, add the acid to water in small quantities, while stirring slowly. Remember the "triple A's"—Always Add Acid to water.

5. Examine the outside of the reagent bottle for any liquid that has dripped down the bottle or spilled on the counter top. Your teacher will show you the proper procedures for cleaning up a chemical spill.

6. Never pour reagents back into stock bottles. At the end of the experiment, your teacher will tell you how to dispose of any excess chemicals.

HEATING MATERIAL IN A TEST TUBE

1. Check to see that the test tube is heat resistant.

2. Always use a test-tube holder or clamp when heating a test tube.

3. Never point a heated test tube at anyone, because the liquid may splash out of the test tube.

4. Never look down into the test tube while heating it.

5. Heat the test tube from the upper portions of the tube downward and continuously move the test tube, as indicated in **Figure H.** Do not heat any one spot on the test tube. Otherwise, a pressure buildup may cause the bottom of the tube to blow out.

USING A MORTAR AND PESTLE

1. A mortar and pestle should be used for grinding only one substance at a time. See **Figure I.**

2. Never use a mortar and pestle for simultaneously mixing different substances.

3. Place the substance to be broken up into the mortar.

4. Firmly push on the pestle to crush the substance. Then grind it to pulverize it.

5. Remove the powdered substance with a porcelain spoon.

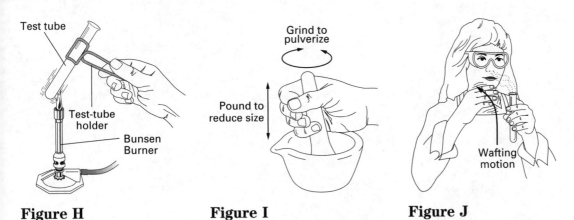

Figure H   **Figure I**   **Figure J**

DETECTING ODORS SAFELY

1. Test for the odor of gases by wafting your hand over the test tube and cautiously sniffing the fumes, as indicated in **Figure J.**

2. Do not inhale any fumes directly.

3. Use a fume hood whenever poisonous or irritating fumes are involved. **Do not** waft and sniff poisonous or irritating fumes.

Specific Heat

Teacher Notes
TIME REQUIRED One 45-minute lab period

SKILLS ACQUIRED
Collecting data
Communicating
Experimenting
Identifying patterns
Inferring
Interpreting
Organizing and analyzing data

RATING
Easy ◄——|—————|—————|—————|——► Hard
　　　　　 1　　 2　　 3　　 4

Teacher Prep–3
Student Set-Up–3
Concept Level–3
Clean Up–2

THE SCIENTIFIC METHOD

Make Observations Students collect calorimetry data using a variety of metals.

Analyze the Results Analysis questions 1 to 9

Draw Conclusions Conclusions questions 10 to 16 and Analysis question 7

Communicate the Results Analysis questions 1, 4, and 7 and Conclusions questions 11, 15, and 16

MATERIALS

Smaller amounts of metals can be used with this lab if less accuracy is acceptable. Rather than reagent-grade metals, try to locate inexpensive sources, such as iron nails, discarded copper wire, and aluminum pellets. Be sure to eliminate any sharp edges.

Once a calorimeter has been made, it can be kept and reused by successive classes and for other experiments.

Tea infusers from kitchen supply stores can be used instead of test tubes to heat the metal samples. They allow quicker and more complete heating of the metal, and the little water that is left on them during the transfer does not appreciably alter the results.

SAFETY CAUTIONS

Safety goggles, gloves, and a lab apron must be worn at all times to provide protection for the eyes and clothing.

If a hot plate is used, review with students precautions to avoid electric shock. Tie back long hair and loose clothing.

Read all safety cautions, and discuss them with your students.

Remind students to use beaker tongs when handling the beaker containing hot water because it can burn or scald.

Remind students of the following safety precautions:

- Always wear safety goggles, gloves, and a lab apron to protect your eyes and clothing. If you get a chemical in your eyes, immediately flush the chemical out at the eyewash station while calling to your teacher. Know the location of the emergency lab shower and the eyewash stations and procedures for using them.

- Do not touch any chemicals. If you get a chemical on your skin or clothing, wash the chemical off at the sink while calling to your teacher. Make sure you carefully read the labels and follow the precautions on all containers of chemicals that you use. If there are no precautions stated on the label, ask your teacher what precautions you should follow. Do not taste any chemicals or items used in the laboratory. Never return leftover chemicals to their original containers; take only small amounts to avoid wasting supplies.

- Call your teacher in the event of a spill. Spills should be cleaned up promptly, according to your teacher's directions.

- Never put broken glass in a regular waste container. Broken glass should be disposed of properly.

- When using a Bunsen burner, confine long hair and loose clothing. If your clothing catches on fire, WALK to the emergency lab shower and use it to put out the fire. Do not heat glassware that is broken, chipped, or cracked. Use tongs or a hot mitt to handle heated glassware and other equipment because hot glassware does not always look hot.

DISPOSAL

Set out three disposal containers, one for each kind of metal. All water can be poured down the drain. The metals should be saved and reused. Be certain that they are well dried before storing them.

TECHNIQUES TO DEMONSTRATE

At this point in the year, students should be familiar with all of the equipment necessary for this lab. Be sure to emphasize that thermometers are too fragile to be used as stirring rods.

Make certain students understand that the more air-tight their calorimeter is, the less energy will escape. Demonstrate how to quickly transfer the hot metal sample from the test tube to the calorimeter without splashing water out of the calorimeter.

TIPS AND TRICKS

Better results can be obtained by repeating the calibration steps to determine an average value for the specific heat of the calorimeter, C', but this may require two lab periods.

Many students have difficulty relating the gain in energy as heat of the water in the calorimeter to the loss in energy as heat of the metal. Refer back to the discussion of the conservation of energy and the flow of energy as heat needed to help them relate this specific idea to a larger context.

Inquiry

Specific Heat

Today's Housewares, Inc. is planning to introduce a new line of cookware. In deciding which metals to use for the cookware, they have to consider how well the metals absorb or release energy as heat. The metals being considered by the company are aluminum, iron, and copper. Your company has a contract with Today's Housewares, Inc. to test the specific heat capacity of each type of metal. This information, along with other considerations, will be used by Today's Housewares, Inc. to decide which metal to use.

BACKGROUND

The amount of energy transferred as heat cannot be measured directly, but can be determined by measuring changes in temperature. To calculate the amount of energy absorbed or lost by a substance as it changes temperature use the following equation.

$$\text{change in heat} = m \times c_p \times \Delta t$$

Metals with high specific heats tend to heat up and cool down slowly because it takes more energy to raise their temperature.

Measurements in a calorimeter are based on the assumption that the energy absorbed by the water in the calorimeter is the same as the energy released as the metal cools down. However, some energy is lost because it is impossible to insulate perfectly. The problem of energy loss is usually solved by calibrating the calorimeter to determine what adjustments are necessary to compensate for the energy that is lost.

PROBLEM

To get the information for your report to Today's Housewares, Inc., you will need to do the following.

- Make and calibrate a calorimeter using carefully measured masses and temperatures of warm and cold water.

- Measure the mass of each metal sample and heat each one, measuring its temperature.

- Add the heated metal to a carefully measured mass and temperature of cold water in the calorimeter.

- Measure the final temperature of the water and metal to calculate the change in energy for the water.

- Determine the specific heat capacity of the metal after accounting for the adjustment specified by the calibration step.

| Specific Heat *continued*

OBJECTIVES

Build and calibrate a simple calorimeter.

Relate measurements of temperature to changes in energy.

Demonstrate proficiency in using calorimetry techniques to determine the specific heat capacities of metals.

MATERIALS

- aluminum metal sample
- balance
- beaker tongs
- beakers, 400 mL (2)
- boiling chips
- copper metal sample
- glass stirring rod
- gloves
- graduated cylinder, 100 mL
- iron metal sample
- lab apron
- plastic foam cups for calorimeter (2)
- ring stand
- safety goggles

- scissors
- test-tube holder
- test tube, large or medium (3)

Bunsen burner option
- Bunsen burner and related equipment
- ring clamp
- wire gauze with ceramic center

Hot plate option
- hot plate

Probe option
- thermistor probe

Thermometer option
- thermometer, nonmercury

Always wear safety goggles, gloves, and a lab apron to protect your eyes and clothing. If you get a chemical in your eyes, immediately flush the chemical out at the eyewash station while calling to your teacher. Know the location of the emergency lab shower and eyewash station and the procedures for using them.

Do not heat glassware that is broken, chipped, or cracked. Use tongs or a hot mitt to handle heated glassware and other equipment because hot glassware does not always look hot.

When using a Bunsen burner, confine long hair and loose clothing. If your clothing catches on fire, WALK to the emergency lab shower and use it to put out the fire.

When heating a substance in a test tube, the mouth of the test tube should point away from where you and others are standing. Watch the test tube at all times to prevent the contents from boiling over.

 Scissors are sharp; use with care to avoid cutting yourself or others.

Specific Heat *continued*

PROCEDURE

Part 1–Preparation

1. Construct a calorimeter from two plastic foam cups. Trim the lip of one cup, and use that cup for the top of the calorimeter. Use a pencil to make a small hole in the center of the base of this cup (the top of the calorimeter) so that a thermometer or thermistor can be inserted, as shown in **Figure 1.** Place the calorimeter in one of the beakers to help prevent it from tipping over.

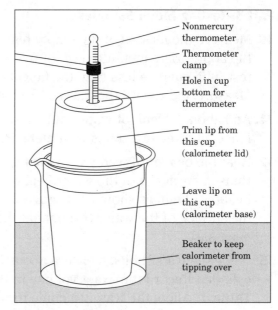

2. Label three test tubes *Al, Cu,* and *Fe.*

Figure 1

Part 2–Calibrating the calorimeter

3. Use a graduated cylinder to measure about 50 mL of cold water. Record the exact volume to the nearest 0.1 mL in **Table 1.** (Note: place an "X" in the box on the *Volume (mL)* row in the column labeled *Calorimeter.* The volume of the calorimeter will not be used in this lab.)

4. Pour the cold water into your calorimeter, and allow the water to reach room temperature. Measure the temperature every minute until the temperature stays the same for 3 min. Record this temperature as the initial temperature for both the cool water and the calorimeter.

5. If you are using a Bunsen burner, clamp a ring to the ring stand, and place the other beaker on the wire gauze on top of the ring clamp. If you are using a hot plate, place the beaker directly on the hot plate. Heat about 75 mL of water in the beaker to 70–80°C.

6. Using beaker tongs, remove the beaker from the heat source, and, using tongs or a hot mitt, pour about 50 mL of hot water into a graduated cylinder. Measure and record the volume to the nearest 0.1 mL in the calibration data table.

7. Using a glass stirring rod, stir the hot water in the graduated cylinder for about a minute and measure the temperature. Record this value in the calibration data table as the initial hot-water temperature. Never use a thermometer to stir anything. It will break easily. The glass wall surrounding the bulb is very thin to provide quick and accurate temperature readings.

8. Immediately pour all of the hot water into the calorimeter with the cool water. Put on the lid and gently swirl the beaker in which the calorimeter rests for 30 s. Record the highest temperature attained by the mixture as the final temperature for the hot water, the cool water, and the calorimeter.

9. Empty the calorimeter. Pour the water down the drain.

Part 3–Testing Metal Samples

10. Measure the masses of the empty test tubes, and record them in **Table 1.** Add the appropriate metal to each of the test tubes. The height of the metal in the test tube must be less than the height of the water in the calorimeter cup. Measure and record the mass of each test tube with metal.

11. Add about 300 mL of water and a few boiling chips to the beaker from **step 5.** Place all three test tubes with metal in the beaker.

12. Heat the beaker of water with the test tubes and metal using a hot plate or a Bunsen burner until the water begins to boil, and continue heating for approximately 10 min as shown in **Figure 2.** Be sure that the metal remains below the surface of the water throughout the heating cycle. Add more water only if necessary.

13. While the water boils, measure about 75 mL of cold water in a graduated cylinder. Record the exact volume to the nearest 0.1 mL in the *Al* column in **Table 1.** Allow the water to come to room temperature, as in **step 4.** Record this value as the initial water temperature in **Table 1.**

14. Measure and record the temperature of the boiling water after 10 min. You can assume that each metal is at the same temperature as the boiling water. Record the value in **Table 1** as the initial temperature for each metal. Keep the water boiling through the following steps.

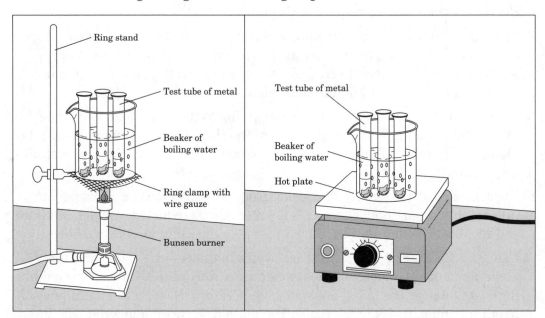

Figure 2

| Specific Heat *continued*

---

15. First make certain that the thermometer is not in the calorimeter because adding the metal could cause it to break. Then remove the *Al* test tube from the hot-water bath with a test-tube holder because it is hot enough to cause burns. Transfer the aluminum metal to the calorimeter. Quickly and carefully put the top on the calorimeter and gently insert the thermometer into the water. Gently swirl the beaker containing the calorimeter for 30 s, making sure that the metal does not hit the thermometer. Monitor the temperature, and record in **Table 1** the highest temperature attained by the water.

16. Remove the aluminum sample from the calorimeter, and pour the water down the drain. Place the aluminum sample on a paper towel. When it is dry, put it in the appropriate disposal container.

17. Repeat **steps 13** and **15–16** for samples of copper and iron.

18. Turn off the boiling water and allow it to cool. Place the dried metals in separate disposal containers designated by your teacher. Clean up your equipment and lab station. After the water has cooled, pour it down the drain. Always wash your hands thoroughly before you leave the lab after cleaning up the lab area and equipment.

TABLE 1 SPECIFIC HEAT DATA

Calibration of Calorimeter			
Measurement	**Cool H_2O**	**Hot H_2O**	**Calorimeter**
Volume (mL)	50.0	50.0	X
Init. temp. (°C)	25.0	75.0	25.0
Final temp. (°C)	49.5	49.5	49.5
Testing of Metals			
Measurement	**Al**	**Cu**	**Fe**
Test-tube mass (g)	28.0	29.5	28.8
Metal + test-tube mass (g)	63.2	116.7	98.0
Metal init. temp. (°C)	99.0	99.0	99.0
H_2O volume (mL)	75.0	75.0	75.0
H_2O init. temp. (°C)	21.0	21.0	21.0
Final temp. (°C)	28.2	28.2	28.1

| Specific Heat *continued*

Analysis

1. **Organizing Ideas** State the scientific law that is the basis for the assumption that the energy lost by the metal as it cools will be equal to the energy gained by the water and the calorimeter.

   The assumption that energy released by one part of the system is absorbed

   by another part of the system is based on the law of conservation of energy,

   which states that energy can be transferred from one object to another or

   converted from one form to another but cannot be created or destroyed.

2. **Organizing Data** Assuming that 1.00 g/mL is the density of cool water and 0.97 g/mL is the density of hot water, calculate the masses of cool and hot water used in the calibration step.

   $$50.0 \text{ mL cool } H_2O \times \frac{1.00 \text{ g}}{1 \text{ mL}} = 50.0 \text{ g cool } H_2O$$

   $$50.0 \text{ mL hot } H_2O \times \frac{0.97 \text{ g}}{1 \text{ mL}} = 48.5 \text{ g hot } H_2O$$

3. **Analyzing Information** Calculate ΔT for the hot water and the cool water.

   $\Delta T_{cool\ H_2O} = 49.5°C - 25.0°C = 24.5°C$
   $\Delta T_{hot\ H_2O} = 75.0°C - 49.5°C = 25.5°C$

4. **Applying Ideas** Using the value of the specific heat capacity of water, which is 4.18 J/g·°C, and the specific heat capacity equation given in the Background section, calculate the amounts of energy as heat released by the hot water and absorbed by the cool water.

   energy absorbed by cool H_2O = (4.18 J/g·°C)(50.0 g)(24.5°C) = 5.12 kJ

   energy released by hot H_2O = (4.18 J/g·°C)(48.5 g)(25.5°C) = 5.17 kJ

5. Evaluating Data Explain why the two values you calculated in item **4** should be slightly different. How much energy as heat was released by the calorimeter to the surroundings?

The values should be slightly different because some energy as heat will be

lost to the surroundings. That is why not all of the energy as heat released

by the hot water was absorbed by the cold water. The energy lost to the

surroundings was 0.05 kJ, or 50 J.

6. Applying Ideas To be able to use a calorimeter for many different temperature changes without calibrating every time, chemists often determine the heat capacity of the calorimeter, or C', from their calibration data, according to the following equation.

$$C' = \frac{\text{heat lost to surroundings}}{\Delta T_{calorimeter}}$$

Determine C' for your plastic foam cup calorimeter.

$$C' = \frac{0.05 \text{ kJ}}{24.5°C} = 2 \text{ J/°C}$$

7. Evaluating Methods Explain whether it is best to have a high or a low value for the heat capacity of a calorimeter.

The lower the value of C', the less energy will escape into the surroundings,

and the more accurate any calculations of energy changes are likely to be.

_____

8. Organizing Data Calculate the mass of each metal used.
   63.2 g − 28.0 g = 35.2 g Al
   116.7 g − 29.5 g = 87.2 g Cu
   98.0 g − 28.8 g = 69.2 g Fe

| Specific Heat *continued*

9. **Organizing Data** Calculate the temperature changes, ΔT, for the water and for the metal in each of the three tests.

Aluminum test

$28.2°C - 21.0°C = 7.2°C = \Delta T_{H_2O}$

$99.0°C - 28.2°C = 70.8°C = \Delta T_{Al}$

Copper test

$28.2°C - 21.0°C = 7.2°C = \Delta T_{H_2O}$

$99.0°C - 28.2°C = 70.8°C = \Delta T_{Cu}$

Iron test

$28.1°C - 21.0°C = 7.1°C = \Delta T_{H_2O}$

$99.0°C - 28.1°C = 70.9°C = \Delta T_{Fe}$

Conclusions

10. **Organizing Ideas** Write a valid equation involving these three quantities: energy released by the metal as it cools, energy absorbed by the water as it warms up, and energy released to the surroundings by the calorimeter. (Hint: refer to item **1**.)

energy released by metal = energy absorbed by H₂O + energy released to surroundings

11. **Applying Ideas** Identify which two of the quantities in the equation from item **10** can be calculated from values you already know, have measured, or have calculated in the previous items. (Hint: apply the specific heat capacity equation and the equation in item **6**.)

energy absorbed by H₂O = $m_{H_2O} \times c_p \times \Delta T_{H_2O}$

energy released to surroundings = $C' \times \Delta T_{H_2O}$

(Note that the change in temperature for the water is assumed to be the

change in temperature for the calorimeter.)

12. **Applying Ideas** Set up three different calculations, one for each of the metals tested, to determine the values of the quantities identified in item **11.** Assume that the density of water is 1.00 g/mL.

Aluminum test

energy absorbed by H_2O = 75.0 g × 4.18 J/g·°C × 7.2°C = 2.3 kJ

energy released to surroundings: 2 J/°C × 7.2°C = 14 J = 0.014 kJ

Copper test

energy absorbed by H_2O = 75.0 g × 4.18 J/g·°C × 7.2°C = 2.3 kJ

energy released to surroundings: 2 J/°C × 7.2°C = 14 J = 0.014 kJ

Iron test

energy absorbed by H_2O = 75.0 g × 4.18 J/g·°C × 7.1°C = 2.2 kJ

energy released to surroundings: 2 J/°C × 7.1°C = 14 J = 0.014 kJ

13. **Inferring Conclusions** Using the specific heat capacity equation and the answers from the previous items, determine the value of the specific heat capacity for each metal.

Aluminum test

energy released by metal = 2.3 kJ + 0.014 kJ = 2.3 kJ, to correct significant figures

2.3 kJ = $c_{p,\ Al}$(35.2 g Al)(70.8°C)

$c_{p,\ Al}$ = 0.92 J/g·°C

Copper test

energy released by metal = 2.3 kJ + 0.014 kJ = 2.3 kJ, to correct significant figures

2.3 kJ = $c_{p,\ Cu}$(87.2 g Cu)(70.8°C)

$c_{p,\ Cu}$ = 0.37 J/g·°C

Iron test

energy released by metal = 2.2 kJ + 0.014 kJ = 2.2 kJ, to correct significant figures

2.2 kJ = $c_{p,\ Fe}$(69.2 g Fe)(70.9°C)

$c_{p,\ Fe}$ = 0.45 J/g·°C

14. **Evaluating Conclusions** Compare your values for specific heats to the values found in a handbook, such as the *CRC Handbook of Chemistry and Physics* or *Lange's Handbook of Chemistry*. (You may need to convert the values given from calories to joules. The conversion factor is: 1 cal = 4.184 J.) Calculate your percent error.

$$\frac{0.92 - 0.90}{0.90} \times 100 = 2\% \text{ for Al}$$

$$\frac{0.37 - 0.38}{0.38} \times 100 = 3\% \text{ for Cu}$$

$$\frac{0.45 - 0.44}{0.44} \times 100 = 2\% \text{ for Fe}$$

15. **Organizing Ideas** Refer to your data to determine how the specific heat capacity of the metals compares to the specific heat capacity of water. Which would absorb the same amount of energy as heat with a greater change in temperature—10 g of metal or 10 g of water?

Metals have low specific heat capacities compared with water. The

temperature of the metals will change more than the temperature of the

water will change.

16. **Evaluating Conclusions** Today's Housewares, Inc. is trying to make pots and pans that will quickly reach cooking temperature. Explain, using your data, which metal is the best choice if this is the primary concern.

Copper is the metal that heats up the quickest because it has the lowest

specific heat capacity.

Extensions

1. **Designing Experiments** Share your data with other lab groups, and calculate a class average for the different specific heat capacities of the metals. Compare the averages to the figures in a chemical handbook, and calculate the percent error for the class averages.

Students' answers will vary depending on class answers. Make sure

calculations of averages are accurate. Usually, the class average is closer

than many of the individual measurements.

2. **Applying Ideas** Design a different calorimeter with better insulation. Describe what tests you would use to calibrate it and measure its heat capacity, C'. If your teacher approves the design of the calorimeter and your plans for testing it, build the calorimeter and determine how much better it works. Compare this value to the value calculated in this lab, and describe the improvement you achieved as a percentage.

Students' suggestions for improving the calorimeter's insulation will vary. Be

sure answers are safe and include carefully planned procedures.

Specific Heat *continued*

---

3. **Applying Ideas** Today's Housewares, Inc. is trying to design the handles for its cookware. Should they select materials with high or low specific heat capacities? Explain your answer.

To prevent burns, cookware handles should be made of substances with high

specific heat capacities, which can absorb considerable energy with small

changes in temperature.

Factors Affecting CO_2 Production in Yeast

Teacher Notes

TIME REQUIRED 120 min if all three parts of the investigation are done concurrently

SKILLS ACQUIRED

Collecting data
Communicating
Designing experiments
Experimenting
Identifying patterns
Inferring
Interpreting
Measuring
Organizing and analyzing data
Predicting

RATING

Easy ◄——1——2——3——4——► Hard

Teacher Prep–2
Student Set-Up–4
Concept Level–2
Clean Up–3

THE SCIENTIFIC METHOD

Make Observations Students observe CO_2 production of yeast under different conditions.

Analyze the Results Analysis questions 1 to 5

Draw Conclusions Conclusions questions 1 to 4

Communicate the Results Analysis questions 1 to 5

MATERIALS

If an incubator is not available, provide students with an insulating material such as bubble wrap or newspaper to wrap the selected flasks to prevent heat loss.

SAFETY CAUTIONS

Read all safety precautions, and discuss them with your students.

• Safety goggles and a lab apron must be worn at all times.

• In case of a spill, use a dampened cloth or paper towels to mop up the spill. Then rinse the towels in running water at the sink, wring them out until only damp, and put them in the trash.

13A Teacher's Guide: Inquiry Experiment

TECHNIQUES TO DEMONSTRATE

You may want to show students how to cover the mouth of a flask with a balloon and how to secure the balloon with a rubber band.

TIPS AND TRICKS

Answers will vary in **Data Table 1,** but the amount of CO_2 produced should increase with time, except for flask *1*, which should produce no CO_2 gas. The amount of CO_2 produced should increase from flask *1* to flask *3*.

Answers will vary in **Data Table 2,** but flask *4* should produce little or no CO_2 gas. The gas volume for flask *6* will be greater than the gas volume for flask *5*.

Answers will vary in **Data Table 3.** The pH values for flask *8* and flask *9* should decrease, and for flask *7*, pH may decrease or remain the same. Flask *8* will have the highest volume of CO_2 gas.

DISPOSAL

Solutions containing yeast and sucrose may be poured down the drain.

Inquiry

Factors Affecting CO$_2$ Production in Yeast

Fermentation is a chemical process in which microorganisms such as bacteria, molds, and yeasts break down energy-rich organic materials, producing energy, alcohol or water, and carbon dioxide gas. Around 1815, Joseph Louis Gay-Lussac, a French chemist, concluded that the production of alcohol and carbon dioxide during fermentation was the result of a physical change in which inert matter decomposed into carbon dioxide and alcohol. This view was accepted until Louis Pasteur proved that fermentation was in fact caused by living microorganisms. Today, fermenters, or bioreactors, are used to manufacture alcoholic beverages, cheese, bread, and pharmaceutical products. In a bioreactor, living cells are mixed with nutrients and grown in a carefully controlled, sterile environment. To make a specific product, the reaction environment's temperature, pressure, pH, oxygen content, and nutrient content must be maintained at levels optimum for the desired product.

In this experiment, you will vary the sugar concentration, temperature, and pH and observe the effect on the natural fermentation process by monitoring and calculating the volume of CO$_2$ gas produced.

OBJECTIVES

Calculate the volume of carbon dioxide gas produced under various conditions.

Graph carbon dioxide production data.

Determine ideal growth conditions for yeast and their relation to carbon dioxide production.

MATERIALS

- sodium carbonate, Na$_2$CO$_3$, 15 g
- *Saccharomyces cerevisiae* (baker's yeast)
- sucrose (table sugar)
- white vinegar
- 250 mL beakers, 3
- thermometer, nonmercury
- balance
- small spherical balloons, 9
- Erlenmeyer flasks or glass bottles, 9
- ice water bath
- incubator or insulating material
- strips of pH paper
- rubber bands, 9
- tape measure
- wax pencil

| Factors Affecting CO_2 Production in Yeast *continued*

  Always wear safety goggles and a lab apron to protect your eyes and clothing. If you get a chemical in your eyes, immediately flush the chemical out at the eyewash station while calling to your teacher. Know the locations of the emergency lab shower and the eyewash station and the procedures for using them.

Do not touch any chemicals. If you get a chemical on your skin or clothing, wash the chemical off at the sink while calling to your teacher. Make sure you carefully read the labels and follow the precautions on all containers of chemicals that you use. If there are no precautions stated on the label, ask your teacher what precautions you should follow. Do not taste any chemicals or items used in the laboratory. Never return leftovers to their original containers; take only small amounts to avoid wasting supplies.

Call your teacher in the event of a spill. Spills should be cleaned up promptly, according to your teacher's directions.

Never put broken glass in a regular waste container. Broken glass should be disposed of properly.

Never stir with a thermometer, because the glass around the bulb is fragile and might break.

Procedure

PART 1: DETERMINING THE EFFECTS OF SUGAR CONCENTRATION ON CO_2 PRODUCTION

1. Use a wax pencil to label nine 250 mL Erlenmeyer flasks or glass bottles *1* to *9*.

2. Add 5 g of dried yeast to each flask.

3. Add no sucrose to flask *1*, 1 g of sucrose to flask *2*, and 5 g of sucrose to flask *3*.

4. To each flask, add 200 mL of lukewarm water (37°C). Gently swirl each flask to mix the contents.

5. Inflate three balloons and allow the air to escape. Repeat this twice.

6. Place the mouth of a deflated spherical balloon over the mouth of each flask, and secure it tightly with a rubber band.

7. Place all the flasks in a culture incubator set at 37°C to maintain the initial temperature. If an incubator is not available, wrap the flask with an insulating material to prevent heat loss.

8. After 30 min, measure the circumference of each balloon. Record your measurements in **Data Table 1.** Measure the balloon's circumference again after 60 min and after 90 min. Record these measurements in **Data Table 1.**

Factors Affecting CO_2 Production in Yeast *continued*

Data Table 1

| Sugar (g) | 30 min | | 60 min | | 90 min | |
	Balloon circumference (cm)	CO_2 volume (mL)	Balloon circumference (cm)	CO_2 volume (mL)	Balloon circumference (cm)	CO_2 volume (mL)
Flask *1*: 0	0	0	0	0	0	0
Flask *2*: 1	11	22	13	37	16	69
Flask *3*: 5	19	120	23	205	27	330

PART 2: DETERMINING THE OPTIMUM GROWTH TEMPERATURE FOR YEAST

9. To each of flasks *4*, *5*, and *6*, add 5 g of sucrose.

10. Add 200 mL of ice water (5°C) to flask *4*, 200 mL of room temperature water (23°C) to flask *5*, and 200 mL of lukewarm water (37°C) to flask *6*. Gently swirl each flask to mix the contents.

11. Inflate three balloons and allow the air to escape. Repeat this twice.

12. Place the mouth of a deflated spherical balloon over the mouth of each flask, and secure it tightly with a rubber band.

13. Place flask *4* in an ice bath and flask *6* in a culture incubator set at 37°C to maintain the initial temperature. If an incubator is not available, wrap flask *6* with an insulating material to prevent heat loss.

14. After 30 min, measure the circumference of each balloon. Record your measurements in **Data Table 2.** Measure the balloon's circumference again after 60 min and after 90 min. Record these measurements in **Data Table 2.**

Data Table 2

| Temp. (°C) | 30 min | | 60 min | | 90 min | |
	Balloon circumference (cm)	CO_2 volume (mL)	Balloon circumference (cm)	CO_2 volume (mL)	Balloon circumference (cm)	CO_2 volume (mL)
Flask *4*: 5	0	0	0	0	0	0
Flask *5*: 23	15	60	19	120	22	180
Flask *6*: 37	19	120	23	205	27	330

| **Factors Affecting CO_2 Production in Yeast** *continued*

PART 3: DETERMINING THE EFFECT OF PH ON CARBON DIOXIDE PRODUCTION

15. To each of flasks *7*, *8*, and *9*, add 5 g of sucrose.

16. To each of three 250 mL beakers, add 200 mL of lukewarm (37°C) water. Label one beaker *pH 2* and add white vinegar drop by drop while stirring until the pH equals 2. Label a second beaker *pH 6* and add, while stirring, white vinegar drop by drop or Na_2CO_3 in small amounts until the pH equals 6. Label the third beaker *pH 10* and slowly add, while stirring, small amounts of Na_2CO_3 until the pH equals 10.

17. Add the solution in beaker *pH 2* to flask *7*. Add the solution in beaker *pH 6* to flask *8*, and add the solution in beaker *pH 10* to flask *9*.

18. Gently swirl each flask to mix the contents.

19. Inflate three balloons and allow the air to escape. Repeat this twice.

20. Place the mouth of a deflated spherical balloon over the mouth of each flask, and secure it tightly with a rubber band.

21. Place the flasks in a culture incubator set at 37°C to maintain the initial temperature. If an incubator is not available, wrap the flask with an insulating material to prevent heat loss.

22. After 30 min, measure the circumference of each balloon. Record your measurements in **Data Table 3.** Measure the balloon's circumference again after 60 min and after 90 min. Record these measurements in **Data Table 3.** After 90 min, test the solutions in each flask with pH paper. Record the results in **Data Table 3.**

Data Table 3

| Initial pH | 30 min | | 60 min | | 90 min | | Final pH |
	Balloon circumference (cm)	CO_2 volume (mL)	Balloon circumference (cm)	CO_2 volume (mL)	Balloon circumference (cm)	CO_2 volume (mL)	
Flask 7: 2	9	12	11	22	14	46	2
Flask 8: 6	19	120	23	205	27	330	4.5
Flask 9: 10	10	17	12	29	13	37	9

DISPOSAL

23. Clean all apparatus and your lab station. Return equipment to its proper place. The solutions containing yeast and sucrose can be poured down the drain. Wash your hands thoroughly after all work is finished and before you leave the lab.

| Factors Affecting CO_2 Production in Yeast *continued*

Analysis

1. **Organizing Data** Calculate the volume of the CO_2 gas in each balloon by using each circumference measurement in **Data Tables 1** through **3** with the following equations. Record your volume calculations in the appropriate tables to the correct number of significant figures.

$$r = \frac{circumference}{2\pi}$$

$$V = \frac{4}{3}\pi r^3$$

Answers will vary. In general, volume should increase from flask *1* to flask *3*, increase from flask *4* to flask *6*, and increase from flask *7* to flask *8*, then decrease from flask *8* to flask *9*. Check that calculations are performed correctly.

2. **Analyzing Results** How is carbon dioxide production dependent on sugar concentration?

As the amount of sugar increases, the amount of CO_2 produced increases.

When no sugar is present, no CO_2 is produced.

_____

_____

3. **Analyzing Results** Based on your results in **Part 2** of the investigation, how does temperature affect CO_2 gas production?

The most CO_2 is produced at 37°C. Less CO_2 is produced at room tempera-

ture, and none is produced at 5°C. Therefore, warmer temperatures favor

fermentation.

_____

_____

Factors Affecting CO$_2$ Production in Yeast *continued*

4. Analyzing Results Compare the initial pH values for each flask in **Part 3** of this investigation with the final pH values. Explain any variation.

Except for flask 2, the final pH is lower than the initial pH. As sugar is

consumed, the CO$_2$ produced reacts with water molecules to form H$_2$CO$_3$,

causing the pH to decrease with time. A pH value of 2 is a bit too acidic, so

there should be little or no CO$_2$ production in flask 7.

_____

_____

5. Evaluating Data Based on your results in **Part 3,** does baker's yeast thrive in an environment that is acidic or basic? Justify your answer.

Most types of yeast thrive in an acidic environment of pH 4 to 6.0. In a

highly acidic environment (pH = 2), CO$_2$ production decreases greatly or

stops altogether. Similarly, a strongly basic environment (pH = 10), impedes

the production of CO$_2$.

_____

_____

Conclusions

1. Predicting Outcomes What would happen to carbon dioxide production if the fermentation reaction were allowed to run overnight?

Carbon dioxide production would continue until the sugar was depleted

because sugar is required for the reaction to take place.

_____

_____

2. Applying Conclusions Carbon dioxide bubbles are responsible for the spongy texture of yeast breads. Based on your lab results, suggest favorable conditions for preparing yeast breads.

Yeasts thrive in lukewarm temperature and slightly acidic pH and with a

sufficient sugar source.

_____

_____

Factors Affecting CO_2 Production in Yeast *continued*

3. **Designing Experiments** Design an experiment that tests the effects of other sugars or sugar substitutes on carbon dioxide production by yeast. Determine the extremes of the temperature range that baker's yeast can withstand.

Answers will vary; sugar substitutes should not react with the yeast.

_____

_____

_____

_____

4. **Analyzing Methods and Designing Experiments** Warmth stimulates the fermentation process. Describe how the procedure in **Part 2** of this experiment could be modified to determine the *ideal* temperature for CO_2 production.

Answers will vary, but procedures should show experiments done over a

range of temperatures from 0°C to 50°C. Above 50°C, the yeast cells die.

_____

_____

Viscosity of Liquids

Teacher Notes

TIME REQUIRED One 45-minute lab period

SKILLS ACQUIRED
Collecting data
Communicating
Experimenting
Identifying patterns
Inferring
Interpreting
Organizing and analyzing data

RATING

Easy ◄—¹——²——³——⁴—► Hard

Teacher Prep–3
Student Set-Up–3
Concept Level–3
Clean Up–3

THE SCIENTIFIC METHOD

Make Observations Students collect calorimetry data using a variety of metals.

Analyze the Results Analysis questions 1 to 9

Draw Conclusions Analysis question 7 and Conclusions questions 10 to 16

Communicate the Results Analysis questions 1, 4, and 7 and Conclusions questions 11, 15, and 16

MATERIALS

For the oil samples, collect the grades of oil indicated in the following table, and label them with the letters *A–F*. It does not matter which type of oil gets labeled with which letter, provided you keep a record. The sample data and calculations shown in these teacher's notes are based on the following table.

Oil type	Letter
SAE-10	C
SAE-20	A
SAE-30	D
SAE-40	F
SAE-50	B
SAE-60	E

The oil samples should be reused from class to class to avoid the need to dispose of them.

Once the pipet viscosimeters have been made and labeled, they can be reused from class to class. For best results, use only one grade of oil in each pipet viscosimeter.

It may help students complete the lab in a single lab period if you provide two sets of samples in test tubes. This way, students can be cooling and heating the different sets at the same time.

SAFETY CAUTIONS

Safety goggles, gloves, and a lab apron must be worn at all times to provide protection for the eyes and clothing.

Tie back long hair and loose clothing.

Read all safety cautions, and discuss them with your students.

Remind students to use a test-tube holder when removing test tubes from the warm-water bath.

The oils used in this lab should not be heated above 60°C.

Remind students of the following safety precautions:

- Always wear safety goggles, gloves, and a lab apron to protect your eyes and clothing. If you get a chemical in your eyes, immediately flush the chemical out at the eyewash station while calling to your teacher. Know the location of the emergency lab shower and the eyewash stations and procedures for using them.

- Do not touch any chemicals. If you get a chemical on your skin or clothing, wash the chemical off at the sink while calling to your teacher. Make sure you carefully read the labels, and follow the precautions on all containers of chemicals that you use. If there are no precautions stated on the label, ask your teacher what precautions you should follow. Do not taste any chemicals or items used in the laboratory. Never return leftovers to their original containers; take only small amounts to avoid wasting supplies.

- Pins are sharp; use them with care to avoid cutting yourself or others.

- When using a Bunsen burner, confine long hair and loose clothing. Do not heat glassware that is broken, chipped, or cracked. Use tongs or a test tube holder to handle heated glassware and other equipment. Because oil tested in this lab is flammable, it should never be heated directly over a flame. Instead use a hot-water bath, and never heat it above 60°C.

- Call your teacher in the event of a spill. Spills should be cleaned up promptly, according to your teacher's directions.

- Never put broken glass in a regular waste container. Broken glass should be disposed of properly.

DISPOSAL

Set out twelve disposal containers, one for each of the six types of oil labeled *A* through *F* and for each of the six different pipets labeled *A* through *F*. Students in successive periods can reuse the labeled beakers, test tubes, and pipets. There is no need for a full-scale cleanup of all equipment until the end of the class day. The oil can be reused class after class and year after year. If the oil is cleaned up with paper towels, they can be disposed of only in a landfill designated for hazardous waste.

TECHNIQUES TO DEMONSTRATE

Be sure to demonstrate exactly how to prepare the pipet viscosimeter. Show students how to control the flow by covering the hole with a finger. It may take several trials before students have a consistent technique that allows measurements to be made. If students have trouble completing the lab in the time allotted, reduce the number of trials run at each temperature.

TIPS AND TRICKS

Make certain that the concepts of intermolecular bonding and hydrocarbon chain length discussed in the textbook are understood. Point out that the larger the molecule, the stronger the London forces can become.

Later, the technique used here of moving the finger away from a hole to allow liquid to flow out of the viscosimeter can be used as an object lesson to help students understand the role of atmospheric pressure. When a finger is held over the hole, the atmospheric pressure outside the pipet holds the fluid inside the stem.

Inquiry

Viscosity of Liquids

SITUATION

You have been contacted by an automotive service shop that received a shipment of bulk containers of motor oil. The containers had been shipped by freight train, but several boxcars had leaky roofs. As a result, the labels peeled off the cans. Before the shop uses this oil in cars, the service technicians must match up the cans with the types of oil that were listed on the shipping invoice, based on the viscosity and the SAE rating of the oils.

BACKGROUND

Viscosity is the measurement of a liquid's resistance to flow. Several factors contribute to viscosity. Liquids with high intermolecular forces tend to be very viscous. For example, glycerol has a high viscosity because of its tendency to form many hydrogen bonds. For other molecules, such as oils, the longer the chain length of the molecule, the more viscous they are. The longer chains not only provide greater surface area for intermolecular attractions, but also can be intertwined more easily. For example, gasoline, which contains molecules that are chains of three to eight carbon atoms, is much less viscous than grease, which usually contains molecules with about 20 to 25 carbon atoms.

The Society of Automotive Engineers rates lubricating oils according to their comparative viscosities. These numerical values, called *SAE ratings*, range from SAE-10 (low viscosity) to SAE-60 (high viscosity) for oils typically used in combustion engines such as those in automobiles and trucks. The ratings are achieved with an instrument called a *viscosimeter*, which has a small capillary tube opening. The amount of time for a specific amount of motor oil to flow through the opening is a measure of viscosity. The less viscous oils flow through in a shorter time than the more viscous oils do.

PROBLEM

To match the correct oil sample to its SAE rating, you will need to do the following.

- Make your own viscosimeter from a pipet.
- Measure the relative viscosities of several oils by timing the oil as it flows through your viscosimeter.
- Measure mass and volume of each oil to calculate density.
- From the measurements, infer which labels belong on the containers of oil.

Viscosity of Liquids *continued*

OBJECTIVES

Demonstrate proficiency in comparing the viscosity of various liquids under identical test conditions.

Construct a small viscosimeter.

Measure flow time of various single-weight oils.

Measure the mass and volume of the oils to calculate density.

Calculate the relative viscosity of the oils.

Graph experimental data.

Compare viscosities and densities to determine the SAE rating of each oil.

MATERIALS

- beakers, 400 mL (2)
- beakers, 50 mL (7)
- distilled water
- gloves
- graduated cylinder, 10 mL
- ice
- lab apron
- oil samples, 10 mL (6)
- pin, straight
- pipets, thin-stem (7)
- ruler, metric
- safety goggles
- stopwatch or clock with second hand
- test-tube holder
- test-tube rack
- test tubes, small (7)
- wax pencil

Bunsen burner option
- Bunsen burner and related equipment
- ring stand and ring
- wire gauze with ceramic center

Hot plate option
- hot plate

Thermometer option
- thermometer, nonmercury
- thermometer clamp

Probe option
- thermistor probe

Always wear safety goggles, gloves, and a lab apron to protect your eyes and clothing. If you get a chemical in your eyes, immediately flush the chemical out at the eyewash station while calling to your teacher. Know the location of the emergency lab shower and eyewash station and the procedures for using them.

Do not touch any chemicals. If you get a chemical on your skin or clothing, wash the chemical off at the sink while calling to your teacher. Make sure you carefully read the labels and follow the precautions on all containers of chemicals that you use. If there are no precautions stated on the label, ask your teacher what precautions to follow. Do not taste any chemicals or items used in the laboratory. Never return leftovers to their original container; take only small amounts to avoid wasting supplies.

| Viscosity of Liquids *continued*

◆ **Do not heat glassware that is broken, chipped, or cracked.** Use tongs or a hot mitt to handle heated glassware and other equipment because hot glassware does not always look hot.

◆ **When using a Bunsen burner, confine long hair and loose clothing.** If your clothing catches on fire, WALK to the emergency lab shower and use it to put out the fire. Because the oil tested in this lab is flammable, it should never be heated directly over a flame. Instead, use a hot-water bath, and never heat it above 60°C.

When heating a substance in a test tube, the mouth of the test tube should point away from where you and others are standing. Watch the test tube at all times to prevent the contents from boiling over.

◆ **Pins are sharp;** use with care to avoid cutting yourself or others.

Procedure

PART 1–PREPARATION

1. Put on safety goggles, gloves, and a lab apron.

2. With a wax pencil, label each 50 mL beaker-test tube-pipet set with the name of one oil sample (*A*, *B*, *C*, *D*, *E*, or *F*). Label an additional set H_2O.

3. Place two marks 2.0 cm apart on the side of the bulb of the pipet, as shown in **Figure 1**. The top mark will be the starting point and the lower mark will be the endpoint.

4. Carefully make a small hole in the top of the bulb of each pipet with the pin, as shown in **Figure 1**. Be sure the hole is well above the marks you made on the side of the pipet bulb. Make the hole the same size for each pipet by putting in the pin the same way for each one. You will control the flow of oil with your finger and this hole.

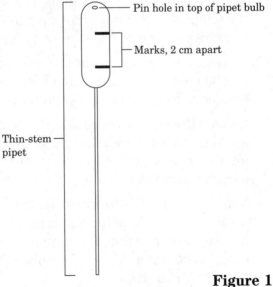

Pin hole in top of pipet bulb

Marks, 2 cm apart

Thin-stem pipet

Figure 1

PART 2–TECHNIQUE

5. Measure the masses of all seven 50 mL beakers. Record them in your data table.

6. Pour about 5.0 mL of distilled water into the graduated cylinder. Measure and record the volume to the nearest 0.1 mL, and pour it into the H_2O beaker.

7. Measure and record the mass of the H_2O beaker with water in your data table.

8. Squeeze the H_2O pipet bulb and fill the pipet with distilled water to above the top line. After it is full, place your finger over the pin hole. Place the pipet over the H_2O beaker, lift your finger off the hole, and allow the liquid to flow into the beaker until the meniscus is even with the top line on the pipet bulb. Cover the hole promptly when the water reaches this point. Several practice trials may be necessary.

9. One member of the lab group should hold the pipet with a finger over the pinhole, and the other should use a clock with a second hand or a stopwatch to record precise time intervals. Hold the pipet over the H_2O beaker. When the timer is ready, remove your finger from the pinhole, and allow the liquid to flow into the beaker until it reaches the bottom line on the pipet bulb. Record the time elapsed to the nearest 0.1 s in your data table in the section for room temperature. (If you do not have a stopwatch, measure the time elapsed to the nearest 0.5 s.) It may take several practice trials to master the technique.

10. Repeat **steps 6–9** with each oil, using the appropriately labeled pipets and beakers. You should perform several trials for each oil and for water to obtain consistent results. Clean the graduated cylinder after the last trial for each oil.

11. Using one of the 400 mL beakers, make an ice bath. Fill the test tubes to within 1.0 cm of the top with the appropriate oil or distilled water. Cool the samples for 5–8 min so that they are at a temperature between 0°C and 10°C. The key is that all of the samples must be at the same temperature. Measure the temperature of the water sample to the nearest 0.1°C with a thermometer or a thermistor probe and record it below your data table.

12. Repeat **steps 8–9** with each of the cooled samples. Be sure to use the pipets and 50 mL beakers designated for each oil or distilled water. Record the volume, mass, and time elapsed for each trial in your data table.

13. Using a Bunsen burner or a hot plate and the second 400 mL beaker, prepare a warm-water bath with a temperature between 35°C and 45°C. If you measure the temperature with a thermometer, use a thermometer clamp attached to a ring stand to hold the thermometer in the water.

14. Refill the test tubes to within 1.0 cm of the top with the appropriate oil or distilled water. Place these test tubes into the warm-water bath, and allow the oil and water to warm. Measure the temperature of the water sample with a thermometer or a thermistor probe when you remove the samples and record it below **Table 1.**

15. Repeat **steps 8–9** with the warm samples. Record the volume, mass, and time elapsed for each trial in your data table.

16. Your instructor will have set out twelve disposal containers; six for the six types of oil and six for the pipets. **Do not pour oil down the sink. Do not put the oil or oily pipets in the trash can.** The distilled water may be poured down the sink. The test tubes should be washed with a mild detergent and rinsed. Always wash your hands thoroughly after cleaning up the area and equipment.

Viscosity of Liquids *continued*

TABLE 1 FLOW TIMES OF THE OILS

Sample	Beaker mass (g)	Total mass (g)	Volume (mL)	Trial 1— cool (s)	Trial 2— cool (s)	Trial 3— cool (s)
A	47.06	51.51	5.0	31.1	31.2	31.3
B	46.81	51.45	5.0	148.5	148.4	148.6
C	47.27	51.63	5.0	20.3	20.2	20.1
D	48.04	52.55	5.0	43.5	43.7	43.6
E	47.53	52.22	5.0	190.5	190.4	190.3
F	47.60	52.15	5.0	74.5	74.6	74.4
H_2O	46.92	51.92	5.0	1.6	1.6	1.6

Sample	Trial 1— room temp. (s)	Trial 2— room temp. (s)	Trial 3— room temp. (s)	Trial 1— warm (s)	Trial 2— warm (s)	Trial 3— warm (s)
A	14.7	14.6	14.5	7.3	7.2	7.1
B	51.3	51.4	51.2	18.3	18.4	18.2
C	10.5	10.3	10.4	5.5	5.3	5.4
D	20.9	21.1	21.0	9.8	9.9	9.7
E	73.6	73.7	73.5	25.5	25.3	25.4
F	31.7	31.6	31.8	11.9	11.7	11.8
H_2O	1.6	1.5	1.6	1.5	1.6	1.5

cool temperature: 4°C

room temperature: 20°C

warm temperature: 40°C

Analysis

1. Organizing Data Determine the density of each sample.

A: $\dfrac{4.45\ g}{5.0\ mL} = 0.89\ g/mL$ E: $\dfrac{4.69\ g}{5.0\ mL} = 0.94\ g/mL$

B: $\dfrac{4.64\ g}{5.0\ mL} = 0.93\ g/mL$ F: $\dfrac{4.55\ g}{5.0\ mL} = 0.91\ g/mL$

C: $\dfrac{4.36\ g}{5.0\ mL} = 0.87\ g/mL$ H_2O: $\dfrac{5.00\ g}{5.0\ mL} = 1.0\ g/mL$

D: $\dfrac{4.51\ g}{5.0\ mL} = 0.90\ g/mL$

Viscosity of Liquids *continued*

2. Organizing Data Find the average flow time for each sample at each temperature.

Sample	Avg. time (s) cool	Avg. time (s) room temp.	Avg. time (s) warm
A	31.2	14.6	7.2
B	148.5	51.3	18.3
C	20.2	10.4	5.4
D	43.6	21.0	9.8
E	190.4	73.6	25.4
F	74.5	31.7	11.8
H_2O	1.6	1.6	1.5

3. Analyzing Information Calculate the relative viscosity of your samples at room temperature by applying the following formula. The values for the absolute viscosity of water are in units of centipoises (cp). A centipoise is equal to 0.01 g/cm·s.

$$\text{relative viscosity}_{oil} = \frac{\text{density}_{oil} \times \text{time elapsed}_{oil} \times \text{viscosity}_{H_2O}}{\text{density}_{H_2O} \times \text{time elapsed}_{H_2O}}$$

Temperature (°C)	Absolute Viscosity for H_2O (cp)
18	1.053
20	1.002
22	0.955
24	0.911
25	0.890
26	0.870
28	0.833

$$\text{relative viscosity}_A: \quad \frac{0.89 \text{ g/mL} \times 14.6 \text{ s} \times 1.002 \text{ cp}}{1.0 \text{ g/mL} \times 1.6 \text{ s}} = 8.1 \text{ cp}$$

$$\text{relative viscosity}_B: \quad \frac{0.93 \text{ g/mL} \times 51.3 \text{ s} \times 1.002 \text{ cp}}{1.0 \text{ g/mL} \times 1.6 \text{ s}} = 30. \text{ cp}$$

$$\text{relative viscosity}_C: \quad \frac{0.87 \text{ g/mL} \times 10.4 \text{ s} \times 1.002 \text{ cp}}{1.0 \text{ g/mL} \times 1.6 \text{ s}} = 5.7 \text{ cp}$$

$$\text{relative viscosity}_D: \quad \frac{0.90 \text{ g/mL} \times 21.0 \text{ s} \times 1.002 \text{ cp}}{1.0 \text{ g/mL} \times 1.6 \text{ s}} = 12 \text{ cp}$$

$$\text{relative viscosity}_E: \quad \frac{0.94 \text{ g/mL} \times 73.6 \text{ s} \times 1.002 \text{ cp}}{1.0 \text{ g/mL} \times 1.6 \text{ s}} = 43 \text{ cp}$$

$$\text{relative viscosity}_F: \quad \frac{0.91 \text{ g/mL} \times 31.7 \text{ s} \times 1.002 \text{ cp}}{1.0 \text{ g/mL} \times 1.6 \text{ s}} = 18 \text{ cp}$$

Conclusions

4. Inferring Conclusions According to the invoice, the service station was supposed to receive equal amounts of SAE-10, SAE-20, SAE-30, SAE-40, SAE-50, and SAE-60 oil. Given that the oils with the lower SAE ratings have lower relative viscosities, infer which oil samples correspond to the SAE ratings indicated.

If the oils were labeled in the table in the Materials section, students should

identify the following matches. Otherwise, refer to your notes about which

oil was labeled with which letter.

Letter	Rel. viscosity (cp)	Oil type
A	8.1	SAE-20
B	30	SAE-50
C	5.7	SAE-10
D	12	SAE-30
E	43	SAE-60
F	18	SAE-40

5. Organizing Information Prepare a graph with flow time at room temperature on the y-axis and SAE rating on the x-axis.

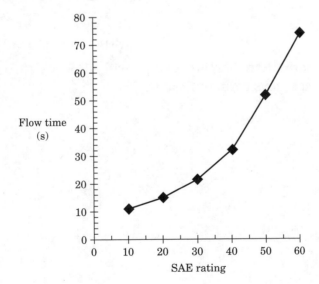

| **Viscosity of Liquids** *continued* |

6. Organizing Information Prepare a graph with density on the *y*-axis and SAE rating on the *x*-axis.

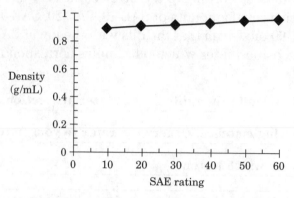

7. Organizing Information Prepare a graph with viscosity at room temperature on the *y*-axis and SAE rating on the *x*-axis.

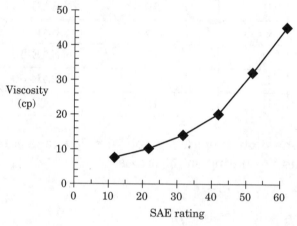

8. Organizing Information Prepare a graph with viscosity at room temperature on the *y*-axis and density on the *x*-axis.

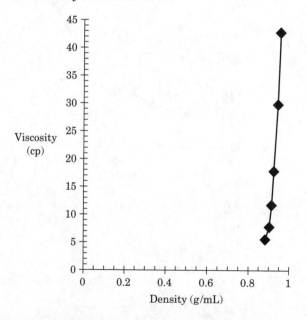

Viscosity of Liquids *continued*

9. Inferring Conclusions How does temperature affect the viscosity of each sample?

All of the oil samples are more viscous at low temperatures and less viscous

at high temperatures.

10. Interpreting Graphics Is there a relationship between density and viscosity?

The densities of the various oils do not vary by much. However, some

students may notice that as the density increases, so does the viscosity.

11. Interpreting Graphics What is the relationship between SAE rating and viscosity?

The lower the SAE rating, the lower the viscosity of the oil.

12. Interpreting Graphics What is the relationship between viscosity and flow time?

More viscous fluids take more time to flow.

Extensions

1. Predicting Outcomes Estimate what flow times you would measure at each temperature if you repeated the tests in this lab with SAE-35 oil.

Given the patterns detected so far, SAE-35 should have flow times between

those of SAE-30 and SAE-40. Answers should be near 59.0 s for 4°C, 26.4 s

for 20°C, and 10.8 s for 40°C.

2. Relating Ideas Malcolm is trying to get the last of the pancake syrup out of a bottle. What can he do to make the syrup come out of the bottle faster? Explain how your plan will take advantage of viscosity.

If Malcolm heats the bottle of pancake syrup, the syrup will be less viscous

and should flow more quickly.

3. Research and Communication Contact a manufacturer of lubrication products such as Valvoline or Pennzoil, and write a short paper on the development and properties of the oils used in this investigation.

Student answers will vary. Be certain students realize that the different

properties of the different grades of oil are a result of the different

combinations of ingredients used for each grade.

Constructing a Heating/Cooling Curve

Teacher Notes

TIME REQUIRED One 45-minute lab period

SKILLS ACQUIRED
Collecting data
Communicating
Experimenting
Identifying patterns
Inferring
Interpreting
Organizing and analyzing data

RATING
Easy ◄——1——2——3——4——► Hard

Teacher Prep–3
Student Set-Up–4
Concept Level–3
Clean Up–3

THE SCIENTIFIC METHOD

Make Observations Students collect temperature data as a substance is cooled and heated, then they create heating and cooling curves.

Analyze the Results Analysis questions 1 to 5 and Conclusions questions 6 and 7

Draw Conclusions Conclusions questions 8 to 10 and Analysis question 4 and 5

Communicate the Results Analysis questions 2 to 4 and Conclusions questions 6 to 10

MATERIALS

This lab will go more quickly if several wire stirrers are prepared in advance. Cut 25 cm lengths of wire. (Any gauge will do provided it is easily bent.) Make a loop at one end that has a 1 cm diameter. Bend the wire where it attaches to the loop so that the loop is perpendicular to the rest of the wire. Bend the top point of the wire over in a small loop to use as a handle.

Fill test tubes with about 15.0 g of $Na_2S_2O_3 \cdot 5H_2O$. The solid can be reused several times. Measure out one 15 g sample, completely transfer it to a test tube, and fill the remaining test tubes to the same level.

Be sure to set out a wide-mouthed bottle containing several small crystals of $Na_2S_2O_3 \cdot 5H_2O$ for students to use as seed crystals.

Other test-tube clamps may be used in place of the three-fingered one.

SAFETY CAUTIONS

Safety goggles, gloves, and a lab apron must be worn at all times.

Read all safety cautions, and discuss them with your students.

Make sure the iron rings are large enough to hold a 600 mL beaker.

Remind students of the following safety precautions:

- Always wear safety goggles, gloves, and a lab apron to protect your eyes and clothing. If you get a chemical in your eyes, immediately flush the chemical out at the eyewash station while calling to your teacher. Know the location of the emergency lab shower and the eyewash stations and procedures for using them.

- Do not touch any chemicals. If you get a chemical on your skin or clothing, wash the chemical off at the sink while calling to your teacher. Make sure you carefully read the labels, and follow the precautions on all containers of chemicals that you use. If there are no precautions stated on the label, ask your teacher what precautions you should follow. Do not taste any chemicals or items used in the laboratory. Never return leftovers to their original containers; take only small amounts to avoid wasting supplies.

- Call your teacher in the event of a spill. Spills should be cleaned up promptly, according to your teacher's directions.

- Never put broken glass in a regular waste container. Broken glass should be disposed of properly.

DISPOSAL

Remelt the sodium thiosulfate pentahydrate in a water bath, pour all of the liquid in a wide-mouth reagent jar, cool to room temperature, cover, and label for reuse. It will be necessary to pulverize the crystals into smaller chunks before reusing them.

TECHNIQUES TO DEMONSTRATE

Students find it particularly awkward to manipulate the setups, so a quick run-through of the steps may prove beneficial: assembling and operating the hot-water bath, positioning the thermometer in the solid, raising and lowering and exchanging the beakers, and when to begin and end timing.

TIPS AND TRICKS

Remind students that thermometer bulbs must not rest on the bottom of test tubes or beakers and that the thermometers must never be used to stir anything. The most accurate temperature readings are made when the thermometer is vertical and the line of sight is horizontal, not angled.

Point out the purpose of the initial quick melt. Emphasize that only one temperature is taken during the quick melt. This provides a rough idea of where the melting and freezing point will occur. In the succeeding tests, more attention should be paid to the temperature readings near this quick-melt temperature, and repetitive readings should be expected.

Explain that the temperature of a pure substance remains constant during a phase change. Review what happens at the particle level during melting and freezing.

NOTE: The $Na_2S_2O_3 \cdot 5H_2O$ will undercool unless a seed crystal is added a few degrees above its freezing temperature. The temperature may still dip one or two degrees below the freezing temperature, but it will rise when the liquid is stirred.

Constructing a Heating/Cooling Curve

Sodium thiosulfate pentahydrate, $Na_2S_2O_3 \cdot 5H_2O$, is produced by a local manufacturing firm and sold nationwide to photography shops, paper processing plants, and textile manufacturers. Purity is one condition of customer satisfaction, so samples of $Na_2S_2O_3 \cdot 5H_2O$ are taken periodically from the production line and tested for purity by an outside testing facility. Your company has been tentatively chosen because your proposal was the only one based on melting and freezing points rather than the more expensive titrations with iodine. To make the contract final, you must convince the manufacturing firm that you can establish accurate standards for comparison.

BACKGROUND

As energy flows from a liquid, its temperature drops. The entropy, or random ordering of its particles, also decreases until a specific ordering of the particles results in a phase change to a solid. If energy is being released or absorbed by a substance remaining at the same temperature, this is evidence that a dramatic change in entropy, such as a phase change, is occurring. Because all of the particles of a pure substance are identical, they all freeze at the same temperature, and the temperature will not change until the phase change is complete. If a substance is impure, the impurities will not lose energy in the same way that the rest of the particles do. Therefore, the freezing point will be somewhat lower, and there will be a range of temperatures instead of a single temperature.

PROBLEM

To evaluate the samples, you will need a heating/cooling curve for pure $Na_2S_2O_3 \cdot 5H_2O$ that you can use as a standard. To create and use this curve, you must do the following.

• Obtain a measured amount of pure $Na_2S_2O_3 \cdot 5H_2O$.

• Melt and freeze the sample, periodically recording the time and temperature.

• Graph the data to determine the melting and freezing points of pure $Na_2S_2O_3 \cdot 5H_2O$.

• Interpret the changes in energy and entropy involved in these phase changes.

• Verify the observed melting point against the accepted melting point found in reference data from two different sources.

• Use the graph to qualitatively determine whether there are impurities in a sample of $Na_2S_2O_3 \cdot 5H_2O$.

Constructing a Heating/Cooling Curve *continued*

OBJECTIVES

Observe the temperature and phase changes of a pure substance.

Measure the time needed for the melting and freezing of a specified amount of substance.

Graph experimental data and determine the melting and freezing points of a pure substance.

Analyze the graph for the relationship between melting point and freezing point.

Identify the relationship between temperature and phase change for a substance.

Infer the relationship between energy and phase changes.

Recognize the effect of an impurity on the melting point of a substance.

Analyze the relationship between energy, entropy, and temperature.

MATERIALS

- balance, centigram
- beaker tongs
- beakers, 600 mL (3)
- chemical reference books
- forceps
- gloves
- graph paper
- hot mitt
- ice
- lab apron
- $Na_2S_2O_3 \cdot 5H_2O$
- plastic washtub
- ring clamps (3)
- ring stands (2)
- ruler
- safety goggles

- stopwatch or clock with a second hand
- test-tube clamp
- test tube, Pyrex, medium
- thermometer clamp
- wire gauze with ceramic center (2)
- wire stirrer

Bunsen burner option
- Bunsen burner
- gas tubing
- striker

Hot plate option
- hot plate

Probe option
- thermistor probes (2)

Thermometer option
- thermometers, nonmercury (2)

Always wear safety goggles, gloves, and a lab apron to protect your eyes and clothing. If you get a chemical in your eyes, immediately flush the chemical out at the eyewash station while calling to your teacher. Know the location of the emergency lab shower and eyewash station and the procedures for using them.

| **Constructing a Heating/Cooling Curve** *continued*

Do not touch any chemicals. If you get a chemical on your skin or clothing, wash the chemical off at the sink while calling to your teacher. Make sure you carefully read the labels and follow the precautions on all containers of chemicals that you use. If there are no precautions stated on the label, ask your teacher what precautions to follow. Do not taste any chemicals or items used in the laboratory. Never return leftovers to their original container; take only small amounts to avoid wasting supplies.

Do not heat glassware that is broken, chipped, or cracked. Use tongs or a hot mitt to handle heated glassware and other equipment because hot glassware does not always look hot.

When using a Bunsen burner, confine long hair and loose clothing. If your clothing catches on fire, WALK to the emergency lab shower and use it to put out the fire.

When heating a substance in a test tube, the mouth of the test tube should point away from where you and others are standing. Watch the test tube at all times to prevent the contents from boiling over.

Procedure

PART 1–PREPARATION

1. Put on safety goggles, gloves, and a lab apron.

2. Fill two 600 mL beakers three-fourths full of tap water.

3. Heat water for a hot-water bath. If you are using a Bunsen burner, attach to a ring stand a ring clamp large enough to hold a 600 mL beaker. Adjust the height of the ring until it is 10 cm above the burner. Cover the ring with wire gauze. Set one 600 mL beaker of water on the gauze. If you are using a hot plate, rest the beaker of water directly on the hot plate.

4. Monitor the temperature of the water with a thermometer or a thermistor probe. Complete **steps 5–8** while the water is heating.

5. Cool the water for a cold-water bath. Fill a small plastic washtub with ice. Form a hole in the ice that is large enough for the second 600 mL beaker. Insert the beaker and pack the ice around it up to the level of the water in the beaker.

6. Bend the piece of wire into the shape of a stirrer, as shown in **Figure 1.** One loop should be narrow enough to fit into the test tube, yet wide enough to easily fit around the thermometer without touching it.

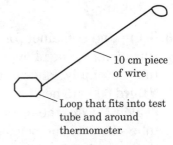

10 cm piece of wire

Loop that fits into test tube and around thermometer

Figure 1

7. Prepare the sample. Assemble the test tube, thermometer, and stirrer, as shown in **Figure 2.** Attach the entire assembly to a second ring stand. Then, add enough $Na_2S_2O_3 \cdot 5H_2O$ crystals so that the test tube is about one-quarter full and the thermometer bulb is well under the surface of the crystals as shown in **Figure 3.**

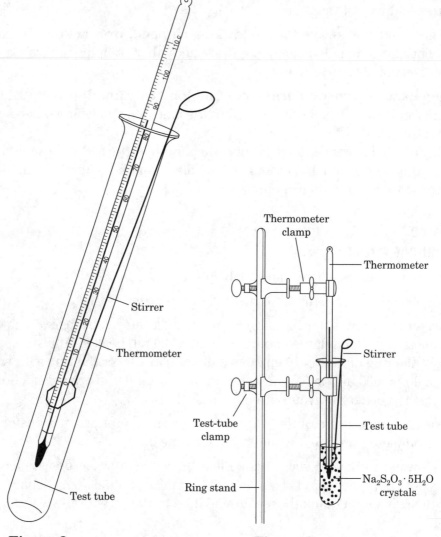

Stirrer

Thermometer

Test tube

Figure 2

Thermometer clamp

Thermometer

Stirrer

Test-tube clamp

Test tube

Ring stand

$Na_2S_2O_3 \cdot 5H_2O$ crystals

Figure 3

8. Set up the container for the hot-water bath as shown in **Figure 4.** Attach two ring clamps, one above the other, to the second ring stand beneath the test-tube assembly. Place a wire gauze with ceramic center on the lower ring. Set a third 600 mL beaker, which should be empty, on the gauze and raise the beaker toward the test-tube assembly until it surrounds nearly one-half of the tube's length. The beaker will pass through the ring clamp without gauze, and the test tube should not touch the bottom or sides of the beaker, as shown in **Figure 4** on the next page. The top clamp keeps the beaker from tipping when the beaker is filled with the hot water.

Constructing a Heating/Cooling Curve *continued*

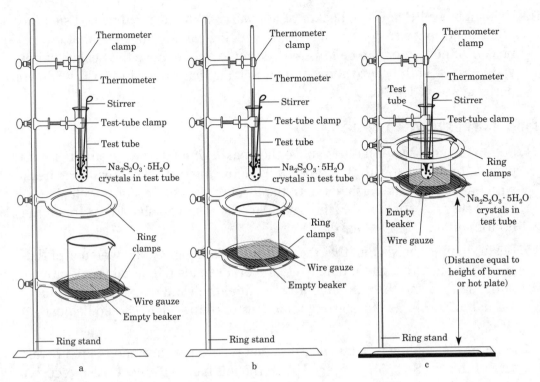

Figure 4

PART 2–MELTING A SOLID: QUICK TEST

9. Check the temperature of the water for the hot-water bath. When it is 85°C, turn off the burner or hot plate. If the temperature is already greater than 85°C, shut off the burner or hot plate, and add a few pieces of ice to bring the temperature down to 85°C. Then, using beaker tongs, remove the beaker of hot water from the burner. Using tongs or a hot mitt carefully pour the water into the empty beaker until the water level is well above the level of the solid inside the test tube. Set the empty beaker on the counter. You will use it again in step **20**.

10. Begin timing. The second the water is poured, one member of the lab group should begin timing, while the other reads the initial temperatures of the bath with one thermometer and sample with the other thermometer.

11. Occasionally stir the melting solid by gently moving the stirrer up and down. Be careful not to break the thermometer bulb. Monitor the temperature of the $Na_2S_2O_3 \cdot 5H_2O$ and the hot-water bath with separate thermometers or probes.

12. When the temperature of the liquid $Na_2S_2O_3 \cdot 5H_2O$ is approximately the same as that of the hot-water bath, stop timing. Note the final temperature of the liquid $Na_2S_2O_3 \cdot 5H_2O$ and the elapsed time. This temperature is the approximate melting point of your sample. Knowing this value can help you make the careful observations necessary to determine a more precise value.

Constructing a Heating/Cooling Curve *continued*

13. Using a hot mitt, hold the beaker of hot water with one hand while using the other hand to gently loosen only the lower ring clamp enough so that the beaker of hot water can be lowered and removed. Remove the beaker of hot water, set it on the gauze above the burner, and let it reheat to 65°C while you perform **steps 14–20.**

PART 3—FREEZING A LIQUID

14. Set up the cold-water bath. Remove the beaker of cold water from the ice and place it on the ring with the gauze, well below the test tube. Steady the beaker with one hand while raising it until the level of the cold water is well above the level of the liquid inside the test tube. The test tube should not touch the bottom or sides of the beaker.

15. Begin timing. The second that the cold water is in place, one member of the lab group should begin timing, while the other reads the initial temperatures of the sample and the bath. Record the initial time and temperatures in the left half of **Table 1.** The starting temperature of the liquid should be near 80°C.

16. Monitor the cooling process. Measure and record the time and the temperature of the $Na_2S_2O_3 \cdot 5H_2O$ every 15 s in the left half of **Table 1.** Also record observations about the substance's appearance and other properties in the *Observations of cooling* column in **Table 1.** When the temperature reaches 50°C, use forceps to add one or two seed crystals of $Na_2S_2O_3 \cdot 5H_2O$ to the test tube.

17. Continue taking temperature readings every 15 s, stirring continuously, until a constant temperature is attained. (A temperature is constant if it is recorded at four consecutive 15 s intervals.) **Do not try to move the thermometer, thermistor probe, or stirrer when solidification occurs.**

18. Finish timing. Continue taking readings until the temperature of the solid differs from the temperature of the cold-water bath by 5°C.

19. Remove the cold-water bath. Grasp the beaker with one hand, carefully loosen its supporting ring clamp with the other hand, and lower the beaker of cold water away from the test tube. Remove the beaker from the ring and set it on the counter.

PART 3—MELTING A SOLID

20. Set up the container for the hot-water bath. Place the empty beaker from **step 9** on the ring and wire gauze. Steady the beaker as you raise it to surround the test tube as you did in **step 8,** but this time allow room for the Bunsen burner to be placed under the beaker.

21. Fill the hot-water bath. Use the second thermometer or thermistor probe to check the temperature of the water for the hot-water bath. When it is 65°C, turn off the burner or hot plate. If the temperature is greater than 65°C, add a few pieces of ice to lower the temperature. Using tongs or a hot mitt, carefully pour the hot water into the empty beaker until the water level is well above the level of the solid inside the test tube. Set the empty beaker on the counter.

22. Begin timing. The second that the water is poured, one member of the lab group should begin timing while another reads initial temperatures of the water bath and the solid $Na_2S_2O_3 \cdot 5H_2O$. Record the solid's temperature in the right half of the data table. The starting temperature of the solid should be below 35°C.

23. Maintain the bath's temperature. Move the burner or hot plate under the hot-water bath and continue heating the water in the bath. Adjust the position and size of the flame or the setting of the hot plate so that the temperature of the hot-water bath remains between 60°C and 65°C.

24. Monitor the warming process. Record the temperature of the sample every 15 s. Use the stirrer, when it becomes free of the solid, to gently stir the contents of the test tube. Also record observations about the substance's appearance and other properties in the *Observations of warming* column in **Table 1.**

25. Continue taking readings until the temperature of the $Na_2S_2O_3 \cdot 5H_2O$ differs from that of the hot-water bath by 5°C.

26. Record the final temperature and the time.

27. Turn off the burner or hot plate.

28. Remove the thermometer or thermistor probe from the liquid $Na_2S_2O_3 \cdot 5H_2O$ and rinse it. Pour the $Na_2S_2O_3 \cdot 5H_2O$ from the test tube into the disposal container designated by your teacher. If you used a Bunsen burner, check to see that the gas valve is completely turned off. Remember to wash your hands thoroughly after cleaning up the lab area and all equipment.

Constructing a Heating/Cooling Curve *continued*

TABLE 1 TIME AND TEMPERATURE DATA

Cooling Data			Warming Data		
Time (s)	Temp. (°C)	Observations of cooling	Time (s)	Temp. (°C)	Observation of warming
0	69.0		0	29.0	
15	61.0		15	31.5	
30	51.0	Seed crystal added	30	36.0	
45	47.0	Crystallization	45	39.5	
60	46.0		60	42.0	
75	48.2		75	43.5	
90	48.2		90	44.5	
105	48.2		105	45.5	
120	48.2		120	46.0	Melting starts
135	48.2		135	46.5	
150	48.2		150	47.0	
165	48.2		165	47.0	
180	48.2		180	47.2	
195	48.2		195	47.4	
210	48.2		210	47.5	
225	48.2		225	47.5	
240	48.0	Total solid	240	47.5	
255	47.5		255	48.0	
270	47.0		270	55.0	All solid melted
285	45.7		285	55.0	
300	44.7		300	58.0	
315	43.5		315	60.0	
330	41.5				
345	39.0				
360	36.2				
375	34.0				
390	32.0				

Analysis

1. **Organizing Data** Plot both the heating and cooling data on the same graph. Place time on the x-axis and temperature on the y-axis.

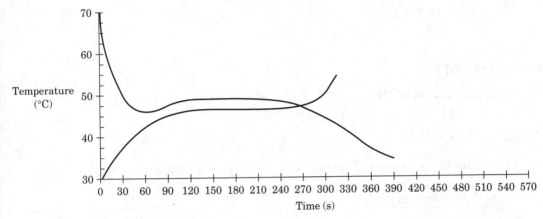

2. **Interpreting Graphics** Describe and compare the shape of the cooling curve with the shape of the heating curve.

The cooling curve begins at an initial high temperature, slopes rapidly

downward to a plateau at 48.2°C, and then continues to slope downward. The

heating curve begins at an initially low temperature, slopes rapidly upward

to a plateau, and then continues to slope upward. The two curves overlap in

the plateau area.

3. **Interpreting Graphics** Locate the freezing and melting temperatures on your graph. Compare them and comment on why they have different names.

The melting point is 48.2°C, while the freezing point is 47.5°C. They are

close but should be the same. The difference in names reflects the phase

change that is taking place and tells from which direction the phase change

is being approached.

4. **Evaluating Methods** One purpose of the quick test for melting point is summarized in **step 12**. State this purpose and explain how it prepares you for **steps 17 and 24**.

The quick melt provides a rough measure of the melting temperature. It tells

the experimenter to pay close attention to time and temperature measure-

ments near this point because the phase change plateau should occur.

| Constructing a Heating/Cooling Curve *continued*

5. Evaluating Data Compare your melting point with that found in references. What is your percent error?

from Merck Index: m.p. = 48°C when rapidly heated

$$\frac{48.2 - 48}{48} \times 100 = 1.67\%$$

Conclusions

6. Analyzing Information As the liquid cools, what is happening to the kinetic energy and the entropy of the following?

a. $Na_2S_2O_3 \cdot 5H_2O$

As the liquid cools, the kinetic energy and entropy decrease for the

$Na_2S_2O_3 \cdot 5H_2O$ **particles.**

b. the water bath

As the liquid cools, the kinetic energy and the entropy of the particles in the

water bath increase.

7. Analyzing Information What happened to the temperature of the sample from the time that freezing began until freezing was complete? Did the entropy of the sample increase, decrease, or stay the same?

The temperature remained relatively constant, but the entropy decreased

while the liquid was freezing into a solid.

_____

8. Predicting Outcomes How would the quantity of the sample affect the time needed for the melting point test?

The larger the sample size, the longer the sample remains at the plateau

temperature.

_____

9. Predicting Outcomes Would the quantity of the sample used to determine the melting point affect its outcome? (Hint: is melting point an extensive or intensive property?)

No, melting point is an intensive property, so it depends on the nature of the

substance being melted, not on the quantity.

_____

Constructing a Heating/Cooling Curve *continued*

10. Interpreting Graphics

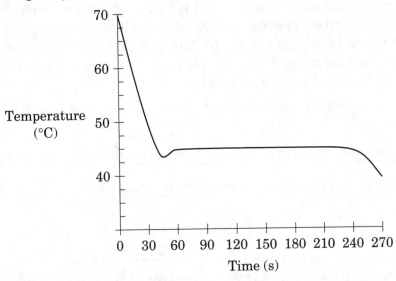

Examine the graph above, and compare it to your cooling curve. Would this sample of sodium thiosulfate pentahydrate be considered pure or impure? Sketch a line on the graph that represents your cooling data. If the curves are not identical, estimate the difference in melting points.

The substance is impure because the melting point is different; ΔT is about

3°C.

Extensions

1. **Interpreting Graphics** Refer to the heating and cooling curves you plotted. For each portion of the curve, describe what happens to the energy and entropy of the substance.

The energy and entropy decrease during cooling but increase during heating.

The downward slope seems to indicate a large transfer of energy to the

surroundings, so molecules are moving less rapidly; the upward slope indi-

cates a large transfer from the surroundings, so molecular motion increases.

The phase changes during plateaus represent times at which there are no

changes in temperature and only small changes in molecular motion, even

though substantial changes occur in energy and entropy.

2. Applying Ideas In northern climates, freezing rain is a driving hazard. When this occurs, warm air from a defroster is blown against the windshield of an automobile in order to restore visibility. It would be convenient to have a system that automatically turned the defroster blower system on and off as needed. A thermostat embedded in the windshield to detect outside temperature could be used to perform this function.

a. At what temperature should the thermostat be set to turn on the hot-air blower?

The thermostat should turn on the freezing-rain defroster when the

temperature is only a few degrees above freezing (0°C).

_____

_____

b. At what temperature should the thermostat be set to turn off the hot-air blower?

The defroster should be turned off when the temperature is warm enough

that the freezing rain should melt immediately, at approximately 20°C.

_____

_____

3. Analyzing Methods Will crystallization take place if no seed crystal is added? Why or why not?

Crystallization should eventually take place when the temperature is below

the freezing point. However, this process will not be as rapid as when the

cooling water bath and seed crystal are used.

_____

_____

4. Designing Experiments Explain the purpose of a water bath. Why wasn't distilled water necessary?

A water bath warms the test tube evenly. Distilled water is not necessary

because the water does not mix with the chemicals.

_____

_____

Constructing a Heating/Cooling Curve *continued*

5. Organizing Ideas Which of the following word equations best represents the changes of phase taking place in the situations described in items **1** to **9** below? Place your answer in the space to the left of the numbered items.

a. solid + energy → liquid

b. liquid → solid + energy

c. solid + energy → vapor

d. vapor → solid + energy

___a___ **1.** ice melting at 0°C

___b___ **2.** water freezing at 0°C

___a, b___ **3.** a mixture of ice and water whose relative amounts remain unchanged

___c___ **4.** a particle escaping from a solid and becoming a vapor particle

___c___ **5.** solids, like camphor and naphthalene, subliming

___a___ **6.** snow melting

___b___ **7.** snow forming

___c___ **8.** dry ice subliming

___d___ **9.** dry ice forming

Testing for Dissolved Oxygen

Teacher Notes

TIME REQUIRED Day 1: 45 min, Day 2: 45 min.

SKILLS ACQUIRED
Collecting data
Communicating
Designing Experiments
Experimenting
Identifying patterns
Inferring
Interpreting
Organizing and analyzing data

RATING

Easy ←————————→ Hard
     1   2   3   4

Teacher Prep–2
Student Set-Up–2
Concept Level–2
Clean Up–1

THE SCIENTIFIC METHOD

Make Observations Students collect data on dissolved oxygen in water at selected temperatures.

Analyze the Results Analysis questions 1 to 3 and Conclusions questions 1 to 4

Draw Conclusions Analysis question 4 and Conclusions questions 1 to 4

Communicate the Results Analysis questions 1 to 3 and Conclusions questions 1 to 4

MATERIALS

Canning jars (8 oz) are acceptable for use in holding the water samples. Be sure the jars you use are heat resistant. If larger jars are used, larger amounts of water than those mentioned in the Procedure section will be necessary.

If your students will be using a dissolved-oxygen meter, be sure to refer to the information accompanying the equipment for detailed operating instructions. It is strongly suggested that a standard solution be used to calibrate the probe or meter in step **10** in the Procedure section. If students share the probe or meter, as many as 15 lab groups can perform this Inquiry with one probe or meter.

Do not use dissolved-oxygen test kits that make use of the Winkler method or other titrimetric methods, because of the dangerous chemicals and the difficulties related to disposal. A simple and less expensive dissolved-oxygen test kit is available from scientific suppliers such as Ward's Natural Science Establishment (1-800-962-2660). The kit contains small reactant-filled ampuls sealed in a vacuum. When the tip of an ampul is broken off and the ampul is dipped into a sample, the sample is drawn inside to react with the reactants, after which it can be compared with standards provided in the kit. When the ampul is used properly, the user will not come into contact with any of the reactants. For this Inquiry, each lab group needs four test ampuls.

SAFETY CAUTIONS

Read all safety cautions, and discuss them with your students.

• Safety goggles and a lab apron must be worn at all times to provide protection for the eyes and clothing.

• Tie back long hair and loose clothing.

• Remind students to use beaker tongs to pour the hot water, because it can burn or scald.

TIPS AND TRICKS

Be certain students understand that solubility is not constant for all conditions but changes when the temperature changes. Explain that to prevent the oxygen content from changing as the samples change in temperature, it is important to keep them in sealed containers that are almost entirely full and free of air. The refrigerated water sample must come to room temperature before testing.

TECHNIQUES TO DEMONSTRATE

Show students how to sample the water while disturbing it as little as possible, to avoid changing the oxygen concentration. Demonstrate the use of the dissolved-oxygen test procedure you will use. If students will use a dissolved-oxygen meter, be sure to demonstrate how it should be calibrated (step **10** of the Procedure section).

If they will be using the dissolved-oxygen test kits, show students how to use them. After they pour a measured sample into the sample cup, they can insert the ampul in one of the depressions in the bottom of the cup. Then, when they press the ampul against the side of the depression, the tip will snap off, and the ampul will fill with water. A small bubble will remain in the ampul. The ampul should be taken out of the cup with the open end downward. The open end should not be sharp, but to prevent injury, students should cover it with tape or a piece of tissue. They should mix the contents of the ampul by inverting it several times, allowing the bubble to travel from end to end. After 2 min, students can compare the color with the standards colors. Show how to use a white background behind the ampul to compare the test color with the standards colors.

DISPOSAL

The water samples may be rinsed down the drain. If you used dissolved-oxygen test kits, as described in the Materials section, collect the used ampuls for further treatment. Because the ampuls contain EDTA, they cannot go to a landfill as is. Put the opened ampuls in an evaporating dish. In a hood known to be operating properly, heat them to at least 250°C for at least 15 min to cause the EDTA to decompose. Ampuls that have been treated in this manner may be discarded in the waste basket. **Never heat sealed ampuls, because they may burst open in a violent manner.**

Testing for Dissolved Oxygen

SITUATION

The company you work for has been hired as an expert witness in a lawsuit. The local chapter of Bass Anglers Unlimited has been disturbed by recent declines in the population of bass in Pulaski Lake. There have been several fish kills, in which large numbers of fish have died at the same time and floated to the surface of the lake, creating a terrible smell and fouling the water. The anglers claim that the fish kills began shortly after the R. C. Throckmorton Power Plant came on line, and they are seeking a court order to shut down the plant. The machinery in the power plant uses water from the lake as a coolant and then returns the water to the lake.

The anglers say that something in the returned water is killing the fish. The utility company operating the power plant points out that they use a closed system that prevents the water from coming into direct contact with the machinery in the plant. They say that the water is just as pure when it comes out as when it goes in; it's just a little warmer. The court has asked you to investigate whether there is a scientific basis for the anglers' claim.

BACKGROUND

Fish rely on the oxygen dissolved in water to live. The water passes through their gills, which remove the oxygen. The less oxygen in the water, the harder the gills have to work to get enough oxygen to keep the fish alive. The normal lake temperature is between 15°C and 17°C. Another witness, a biology professor specializing in ichthyology (the study of fish), has testified that if the oxygen content of the water dips below 90% of the normal value at these temperatures, the fish will suffer long-term damage or death. The temperature of the water returned to the lake from the power plant is 28°C. To measure the change in dissolved-oxygen content over this temperature range, you can use tap water, which has a mineral content similar to that of lake water.

PROBLEM

In order to evaluate these claims, you will need to do the following:

• Prepare water samples at several different temperatures.

• Measure the dissolved-oxygen content in parts per million, ppm, for each one.

• Graph the relationship between temperature and the solubility of dissolved oxygen.

• Extrapolate to determine the solubility of dissolved oxygen at 16°C and 28°C.

• Compare the solubilities to determine if there is enough difference to cause damage to the fish.

OBJECTIVES

Measure the concentration of oxygen in a sample of water.

Graph the relationship between the concentration of a gas and temperature.

Infer a general rule of thumb for gas solubilities and temperature.

Relate changes in gas solubility to a fish kill.

MATERIALS

- ice, about 200 g
- 600 mL beaker
- beaker tongs
- hot mitt
- jars with screw-on lids, 4
- Bunsen burner
- gas tubing
- ring stand
- ring clamp

- sparker
- wire gauze with ceramic center
- hot plate (optional)
- thermistor probe (optional)

Alternative option

- dissolved-oxygen test kit with 4 test ampuls, or dissolved-oxygen meter
- thermometer, nonmercury

Always wear safety goggles and a lab apron to protect your eyes and clothing. If you get a chemical in your eyes, immediately flush the chemical out at the eyewash station while calling to your teacher. Know the locations of the emergency lab shower and eyewash station and the procedure for using them.

When you use a Bunsen burner, confine any long hair and loose clothing. Do not heat glassware that is broken, chipped, or cracked. Use tongs or a hot mitt to handle heated glassware and other equipment because hot glassware does not look hot. If your clothing catches on fire, WALK to the emergency lab shower, and use it to put out the fire.

Always clean up the lab and all equipment after use, and dispose of substances according to proper disposal methods. Wash your hands thoroughly before you leave the lab after all lab work is finished.

Procedure

PART 1: PREPARATION

1. Use the **Data Table** to record your results.

2. Label four jars *ice water, room temp, 50°C,* and *100°C.*

PART 2: SAMPLE PREPARATION

3. Add approximately 50 g of ice to 100 mL of tap water in the *ice water* jar. Let the ice water mixture stand for 5 min. Measure its temperature with a thermometer or a thermistor probe to the nearest 0.1°C. The temperature should

Testing for Dissolved Oxygen *continued*

Data Table		
Water	**Temperature (°C)**	**Dissolved O$_2$ (ppm)**
ice water	5.1	10.0
room temp.	25.2	8.5
50°C	50.0	5.0
100°C	100.0	1.0

be near 4°C. Record the temperature in the **Data Table.** Add more ice until the jar is filled to the rim. Screw on the lid and place the sample in an ice chest or refrigerator overnight.

4. Fill the *room temp* jar with tap water. Leave the jar open. Let it sit out overnight where it will not be disturbed. You will measure the temperature of this sample tomorrow.

5. If you are using a Bunsen burner, set up the ring stand, ring clamp, and wire gauze over the burner so that they will hold a beaker. If you are using a hot plate, continue with step **6.**

6. Pour approximately 450 mL of tap water into a 600 mL beaker, and gently heat it to about 50°C. Maintain 50°C as closely as possible for 5 min. Measure the temperature of the water with a thermometer or a thermistor probe to the nearest 0.1°C. Record the temperature in the **Data Table.** Using beaker tongs to hold the hot beaker, carefully fill the jar labeled *50°C* with the warm water. Fill the container to the rim. Screw on the lid, and store the jar in a safe place.

7. Heat the remaining water to boiling (approximately 100°C). Allow the water to boil for about 30 min. Measure the temperature of the water with a thermometer or a thermistor probe to the nearest 0.1°C. Record the temperature in the **Data Table.** Using beaker tongs to hold the hot beaker, carefully fill the jar labeled *100°C* with the boiling water. Fill the container to the rim. Using a hot mitt, screw on the lid and store the jar in a safe place.

PART 3: SAMPLE TESTING

8. On the following day, retrieve your samples. Allow the refrigerated water sample to come to room temperature with the lid on.

9. Measure the temperature of the water in the *room temp* jar with a thermometer or a thermistor probe to the nearest 0.1°C. Be sure to disturb the water as little as possible. Record the temperature in the **Data Table.**

10. If you are using a dissolved-oxygen probe or meter, it may be necessary to calibrate it using a standardized solution. Ask your teacher for instructions. If your teacher indicates that this is unnecessary, continue with step **11.**

11. Disturbing the water as little as possible, measure the dissolved-oxygen content of each sample, and record the value to the nearest 0.1 ppm in the **Data Table.** If you are using a dissolved-oxygen test kit, you will be comparing

the colors of standard solutions to the colors of your tested solutions. Estimate
the measured concentrations to the nearest 0.5 ppm.

DISPOSAL

12. If you used chemical test kits, dispose of the ampuls and reactants in
the containers designated by your teacher. Rinse the samples down the
drain. Clean all equipment.

Analysis

1. **Organizing Data** Make a graph of your data, with temperature (in °C) plotted
on the horizontal axis and O_2 concentration (in ppm) plotted on the vertical
axis. Draw a straight line that best fits the data.

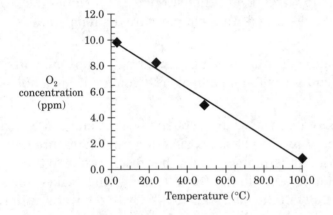

2. **Interpreting Graphics** Refer to the graph from item **1**, and explain in your
own words the relationship between temperature and the solubility of oxygen
in water.

As the temperature increases, the solubility of oxygen decreases.

3. **Analyzing Methods** Why was it important to disturb the water as little as
possible when measuring the dissolved-oxygen content?

If the water had been disturbed, bubbles could have trapped some oxygen.

4. **Applying Models** Often, when water is heated, small bubbles appear long
before boiling begins. Explain why these bubbles form. (Hint: Use the concept
of solubility in your explanation.)

As temperature increases, the amount of gas that the water can hold

decreases. Any amount of gas that is no longer soluble forms bubbles.

Conclusions

1. **Interpreting Graphics** Using the graph from Analysis item **1**, interpolate to determine the concentration of O_2 in the water at 16°C and at 28°C.

   From the graph, the values appear to be about 9.0 ppm for 16°C and 7.9 ppm

   for 28°C.

2. **Interpreting Graphics** Using the format shown below, determine the equation for the line you drew on the graph in Analysis item **1**. (Hint: If you have a graphing calculator, use the STAT mode to enter your data and make a linear regression equation using the LinReg function from the STAT menu.)

   $$O_2 \text{ concentration} = m(\text{temp}) + b$$

   equation from the graphing calculator:

   $$O_2 \text{ concentration} = (-0.095)(\text{temp}) + 10.4$$

3. **Analyzing Conclusions** Calculate the expected concentration of O_2 in the water at 16°C and at 28°C, using the equation you just determined.

   According to the equation, O_2 concentration is 8.9 ppm at 16°C and 7.7 ppm

   at 28°C.

4. **Evaluating Conclusions** Using the values from Conclusions item **1** as the accepted values, calculate the percentage error for the values you calculated in item **3**.

   $$\frac{9.0 - 8.9}{9.0} \times 100 = 1.1\% \text{ error}$$

   $$\frac{7.9 - 7.7}{7.9} \times 100 = 2.5\% \text{ error}$$

Testing for Dissolved Oxygen *continued*

5. Evaluating Conclusions Earlier in the trial, the ichthyologist testified that fish receiving less than 90% of the oxygen normally available at 16°C would suffer long-term damage. From Conclusions item **1** and item **3,** determine whether 90% of the oxygen available at 16°C is available at 28°C. Is it likely that the water is damaging to the fish?

graph answer: 9.0 ppm $\times \dfrac{90}{100}$ = 8.1 ppm = O_2 available at 16°C

equation answer: 8.9 ppm $\times \dfrac{90}{100}$ = 8.0 ppm = O_2 available at 16°C

The O_2 concentration at 28°C is around 7.8 ppm. Student answers will vary,

but they should show that the warm water does not have enough oxygen for

fish.

Extensions

1. Evaluating Conclusions During cross-examination, the attorney for the power plant suggests that, because you tested tap water instead of water taken directly from the power plant, your solubility results are irrelevant and the power plant's water could have even more oxygen in it than the original cold water. Is the attorney correct? Why or why not?

The attorney is incorrect because the solubility will follow the same general

pattern for all samples of water, whether taken from the tap or not.

2. Evaluating Conclusions Later in the cross-examination, the attorney for the power plant asks whether your results establish conclusively that fish died because of the warmer water and for no other reasons. What do you say?

These tests do not conclusively establish that the fish kills were due only

and solely to the water. They establish only that this warm water could be

harmful to fish.

Testing for Dissolved Oxygen *continued*

3. **Designing Experiments** Assuming that you had unlimited laboratory resources and access to all of the lake, what other tests would you perform to be more certain of your results?

   Student answers may vary, but be sure that they are practical and will help

   isolate the cause of the fish kill. Students may suggest testing the water at

   the plant and in the lake and dissecting dead fish to discover the cause of

   death.

   _____

4. **Evaluating Methods** The judge has asked for the opinions of all expert witnesses about a proposed settlement. The power plant proposes to insert a device to bubble oxygen through the warm water as it is released into the lake. Will this solve the problem? Explain why or why not, using the principles of solubility.

   This will not work, because the water will not be able to keep this additional

   oxygen dissolved at the warmer temperatures, and it will simply bubble

   away.

   _____

5. **Relating Ideas** The power plant supplies thousands of households with their electricity. The judge has asked for possible solutions that will keep the power plant working and prevent further damage to the fish. What do you suggest?

   Student answers will vary, but they should indicate some way to restore the

   water to its original temperature or not to return it to the lake at all. Some

   students may suggest pouring the coolant water into containment ponds or

   tanks to cool off before being returned or recirculated.

   _____

Paper Chromatography

Teacher Notes

TIME REQUIRED One 45-minute lab period

SKILLS ACQUIRED
Collecting data
Communicating
Experimenting
Identifying patterns
Inferring
Interpreting
Organizing and analyzing data

RATING
Easy ◄—— 1   2   3   4 ——► Hard

Teacher Prep–2
Student Set-Up–2
Concept Level–3
Clean Up–2

THE SCIENTIFIC METHOD

Make Observations Students perform chromatography experiments in order to separate inks into their component colors.

Analyze the Results Analysis questions 2 to 4, 6, and 7 and Conclusions questions 8 and 9

Draw Conclusions Conclusions questions 8 and 9

Communicate the Results Analysis questions 3, 4, and 7 and Conclusions questions 9 and 10

MATERIALS

Label the pens *1–6*.

Prepare the signatures before students enter the room. First cut a piece of filter paper in half. On each half, about 1.0 cm away from the flat edge, sign "A. Lincoln" with one of the six black ink pens. Be sure to keep track of which pen you use. Note that students will achieve best results if they use two segments of the signature that include fairly flat lines, rather than using the entire signature, or parts with vertical lines or loops.

Figure 1

SAFETY CAUTIONS

The isopropanol is extremely flammable. It should be kept in a closed bottle in an operating fume hood. Place only 300 mL at a time in the bottle. Students should replace the lid when they are finished.

No burners, flames, hot plates, or other heat sources should be in use in the lab when isopropanol is being used.

Remind students of the following safety precautions:

- Always wear safety goggles, gloves, and a lab apron to protect your eyes and clothing. If you get a chemical in your eyes, immediately flush the chemical out at the eyewash station while calling to your teacher. Know the location of the emergency lab shower and the eyewash stations and procedures for using them.

- Do not touch any chemicals. If you get a chemical on your skin or clothing, wash the chemical off at the sink while calling to your teacher. Make sure you carefully read the labels and follow the precautions on all containers of chemicals that you use. If there are no precautions stated on the label, ask your teacher what precautions you should follow. Do not taste any chemicals or items used in the laboratory. Never return leftovers to their original containers; take only small amounts to avoid wasting supplies.

- Call your teacher in the event of a spill. Spills should be cleaned up promptly, according to your teacher's directions.

- Never put broken glass in a regular waste container. Broken glass should be disposed of properly.

DISPOSAL

Set out a disposal container for any isopropanol left over at the end of the procedure. Dilute it with 10 times as much water, and pour it down the drain. Students are instructed to pour the water down the drain. The chromatograms may be discarded in the trash can.

TECHNIQUES TO DEMONSTRATE

Students may need to see an actual chromatography setup before they understand exactly what they will be trying to achieve. This is especially true for the chromatogram of the forged signature. Students need to avoid parts of the signature with loops or vertical lines because these will not make as clear a chromatogram as dots and horizontal lines.

TIPS AND TRICKS

Although this is a popular experiment for students of all ages, for it to be instructionally useful, students must understand that the solubility properties can explain, in part, *why* this particular technique works. Make certain students understand this link. It is important that students realize that the writing sample may take slightly longer to form a chromatogram than did the six dots placed on the filter paper during the lab. If students make measurements of their chromatograms (how far components traveled in what time), this can be made into a more quantitative exercise.

Inquiry

Paper Chromatography

SITUATION

Recently, handwriting experts discovered a set of forgeries. Several museums in the United States had been displaying documents supposedly signed by Abraham Lincoln. The FBI suspects that this could be the work of Benny "Fingers" Smithson, who was recently paroled from prison after serving time for his part in a phony Babe Ruth autograph scam. Smithson denies his involvement. A search warrant was issued, and the FBI found three pens in his apartment. The FBI is also investigating another suspect, Thomas Banks, an employee of one of the museums. Three pens from his belongings are also being held as evidence. Before the FBI can press charges, they need conclusive evidence linking the pens and the phony signatures.

BACKGROUND

Paper chromatography is a method of separating mixtures by using a piece of absorbent paper. In this process, the solution to be separated is placed on a piece of dry filter paper (the stationary phase). A solvent (the moving phase) is allowed to travel across the paper by capillary action. As the solvent is soaked up by the paper, some of the components of the mixture are carried with it. The components of the mixture that are most soluble in the solvent and least attracted to the paper travel the farthest. The resulting pattern of molecules is called a *chromatogram*. In cases where the molecules are easily visible, such as in inks, this method distinguishes the components of a mixture.

PROBLEM

To determine which pen was used in the forgery, you must do the following.

- Prepare chromatograms for each of the pens using the two different solvents.
- Prepare chromatograms for different parts of the forged signature using two different solvents.
- Compare the chromatograms and decide which pen is the likeliest match.
- Provide specific examples of similarities between chromatograms, citing measurable points of comparison.

OBJECTIVES

Demonstrate proficiency in qualitatively separating mixtures using paper chromatography.

Compare inks by using paper chromatography with a variety of solvents.

Evaluate samples to establish which pen was used on a document.

| Paper Chromatography *continued*

MATERIALS

- distilled water
- filter paper wicks, equilateral triangles, 2 cm (2)
- filter papers, 12 cm (4)
- forged signature samples
- gloves
- isopropanol

- lab apron
- paper clips (2)
- pencil
- pens, black ink (6)
- Petri dish with lid
- ruler
- safety goggles

Always wear safety goggles, gloves, and a lab apron to protect your eyes and clothing. If you get a chemical in your eyes, immediately flush the chemical out at the eyewash station while calling to your teacher. Know the location of the emergency lab shower and eyewash station and the procedures for using them.

Do not touch any chemicals. If you get a chemical on your skin or clothing, wash the chemical off at the sink while calling to your teacher. Make sure you carefully read the labels and follow the precautions on all containers of chemicals that you use. If there are no precautions stated on the label, ask your teacher what precautions to follow. Do not taste any chemicals or items used in the laboratory. Never return leftovers to their original container; take only small amounts to avoid wasting supplies.

Isopropanol is flammable. When working with flammable liquids, be sure that no one else in the lab is using a lit Bunsen burner or plans to use one. Make sure that no other heat sources are present. Carry out all work with isopropanol in a hood.

Procedure

1. Put on safety goggles, gloves, and a lab apron.

2. Use a pencil to sketch a circle about the size of a quarter in the center of the piece of filter paper. Write the numbers 1–6 in pencil around the inside of this circle, as shown in **Figure 1** on the next page.

3. On the circle beside the number 1, use pen number 1 to make a large dot. Use pen number 2 to make a dot beside number 2, and repeat this procedure for each pen.

4. Repeat **steps 2 and 3** with a second piece of filter paper. One will be used with water as a solvent, and the second will be used with isopropanol as a solvent.

5. Roll up the triangle of filter paper to be used as a wick. Use the pencil to poke a small hole in the center of the first marked piece of filter paper. Insert a rolled-up piece of the wick through the hole, as shown in **Figure 2** on the next page.

Paper Chromatography *continued*

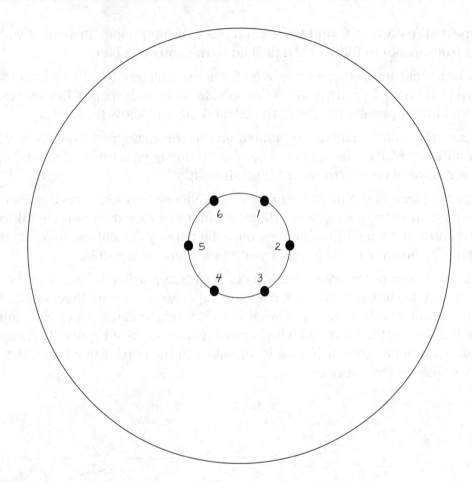

Figure 1

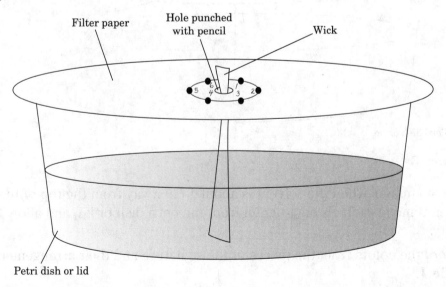

Filter paper

Hole punched
with pencil

Wick

Petri dish or lid

Figure 2

6. Fill the Petri dish to the halfway point with water. Set the wick of the filter
paper into this water, as shown in the illustration, and wait for the chromatogram to develop.

| Paper Chromatography *continued*

7. Repeat **steps 5 and 6** with the second piece of filter paper. Instead of water, use isopropanol to fill the Petri dish lid to the halfway point.

8. Allow the chromatograms to develop for approximately 15 min, or until the solvent is about 1 cm away from the outside edge of the paper. Remove each piece of filter paper from the Petri dish and lid, and allow them to dry.

9. Record the colors that have separated on the chromatogram from each of the six different black inks in your data table. You may either describe the colors or use colored pencils to record this information.

10. Take the piece of the forged signature, and choose two segments that can be used to make chromatograms. (Hint: consider in which direction the ink will travel after the wick brings solvent onto the paper.) Cut out the two segments, leaving as much blank paper attached to each one as possible.

11. With one piece of the paper, make a chromatogram using the water in the Petri dish. Do not allow the ink from the paper to come into direct contact with the solvent. Instead, use a wick as before. Unbend the paper clip, and use it to support the strip with the signature, as shown in **Figure 3**. Then make a chromatogram using the isopropanol in the Petri dish lid with the other piece of the paper.

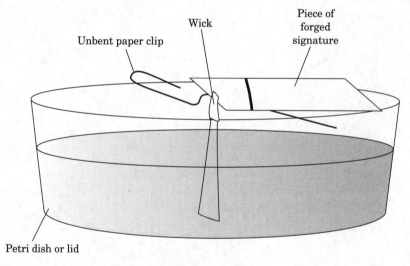

Unbent paper clip   Wick   Piece of forged signature

Petri dish or lid

Figure 3

12. After 15 min or when the solvent is about 1 cm away from the top of the paper, remove each chromatogram from the petri dish or lid, and allow them to dry.

13. Record the colors from the pieces of the signature and their arrangement in **Table 1**.

14. The water may be poured down the sink. Chromatograms and other pieces of filter paper may be discarded in the trash can. The isopropanol solution should be placed in the waste disposal container designated by your teacher. Clean up your equipment and lab station. Thoroughly wash your hands after completing the lab session and cleanup.

TABLE 1 RESULTS OF THE CHROMATOGRAMS

Pen number	H_2O center	H_2O middle	H_2O edge	Isopropanol center	Isopropanol middle	Isopropanol edge
Smithson 1						
Smithson 2						
Smithson 3						
Banks 4						
Banks 5						
Banks 6						
Forgery						

Analysis

1. **Organizing Ideas** Draw Lewis dot structures for water, H_2O, and isopropanol, $CH_3CHOHCH_3$.

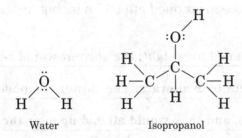

Water   Isopropanol

2. **Analyzing Ideas** Analyze the bonding and structure of water and isopropanol molecules. What types of substances, polar or nonpolar, ionic or covalent, are most likely to dissolve in each one?

Water is a polar molecule that will dissolve polar and charged substances.

Isopropanol has a small polar region but is mostly nonpolar, so it will

dissolve nonpolar substances.

3. **Applying Ideas** Some components of ink are minimally attracted to the stationary phase and very soluble in the solvent. Where on the filter paper should these components be located in the final chromatogram?

The more soluble a component is, the farther it will travel with the solvent.

_____

_____

_____

4. **Relating Ideas** The diffusion of a solute in a solvent is somewhat similar to the diffusion of gases. Where on the filter paper do you expect the larger molecules to be located in the final chromatogram?

   Larger molecules are likely to move more slowly through the solvent and do

   not travel as far as small molecules.

   _____

5. **Evaluating Methods** Explain why the labels numbering the pen spots were written in pencil. (Hint: recall what you know about the nature of the bonds and structure in the graphite that is found in pencils.)

   The graphite in pencil lead is composed of sheets of covalently bonded

   molecules that are nonpolar and insoluble in water. Thus, they will not

   contaminate the chromatogram.

6. **Predicting Outcomes** What would have happened to the chromatogram if the process had not been stopped after 15 min, but instead was allowed to proceed overnight?

   If the process continued overnight, the solvent would reach the edge of the

   filter paper and begin to evaporate. The slower components would catch up

   with the faster ones, and they would all end up near the edge of the filter

   paper.

7. **Predicting Outcomes** At Bryan High School, Armando was performing the chromatogram experiment. He thought it was going too slow and decided to work without a wick. He dipped the sample into the solvent so that the ink dot was even with the top of the water. But when he finished, the chromatogram was too faint to be seen. Explain why Armando's experiment failed.

   When the ink was dipped into the solvent, most of it dissolved in the solvent

   in the dish instead of being pulled with the solvent across the filter paper.

   _____

Conclusions

8. Analyzing Conclusions Are the properties of the component that traveled the furthest in the water chromatogram likely to be similar to the properties of the component that traveled the furthest in the isopropanol chromatogram? Explain your reasoning.

The component that travels the farthest in the water chromatogram is likely

to be small and ionic or polar. The component that travels the farthest in the

isopropanol chromatogram is also likely to be small, but probably nonpolar.

9. Analyzing Conclusions Which pen was used for the forged signature? Explain your reasoning, giving specific examples and providing quantitative data as possible. (Hint: measure the distances that each component traveled to gather quantitative data.)

Student answers will vary, depending on which pen you chose for the forged

signature. Students should justify their choices by pointing out specific

similarities between the forgery chromatograms and the chromatograms of

the pen they chose.

Extensions

1. Designing Experiments How would you improve the efficiency of the separation? If you can think of a way to make the technique work better, ask your teacher to approve your plan, and run the procedure again.

Student suggestions for improving the separation will vary. Some may sug-

gest using a longer piece of filter paper or trying other solvents. Be certain

answers are safe and include carefully planned procedures before allowing

students to proceed.

Paper Chromatography *continued*

2. **Predicting Outcomes** Amino acids are organic chemicals that are monomers of proteins. They each have at least one amino group and one carboxylic acid group, each of which can carry charge in certain solutions. A scientist is analyzing a sample that contains a mixture of the four amino acids shown. She can detect the presence of any amino acid by spraying it with an indicator called *ninhydrin*. She makes two chromatograms, one with water and another with isopropanol, using different portions of the sample. In what order would the amino acids be after each chromatogram was dried and sprayed with ninhydrin? (Hint: it may be helpful to refer to your answers to items **4** and **8**. Also consider what you know about solubility rules.)

Water	Isopropanol
aspartic acid	glycine (or valine)
glycine	valine (or glycine)
valine	phenylalanine
phenylalanine	aspartic acid

The charge on aspartic acid will make it very soluble in water. The next most soluble is likely to be glycine because it is also mostly polar. The other amino acids are both mostly nonpolar, so size should be the dominant factor. In isopropanol, depending on whether size or polarity is the most important factor, glycine or valine will go first. Phenylalanine will dissolve well because it is nonpolar, but it is so much larger that it is likely to lag behind the first two. Lastly, the charges on aspartic acid will decrease its solubility in the nonpolar solvent.

3. Applying Ideas In a gel chromatography apparatus, a polymeric gel that has many pores, or openings, is packed into a long tube or column. A solvent is continually forced under low pressure through the tube. The gel serves as the stationary phase, and the solvent is the mobile phase. A sample of a mixture can be injected into the solvent just before it enters the column, and it will be separated into its component parts. Each component of the sample will take a different amount of time to pass through the column. If the solvent used is benzene, in what order will the plant pigments shown below pass through the column? (Hint: it may be helpful to refer to your answers to items **4** and **8**. Also consider what you know about solubility rules.)

Beta carotene (yellow/orange)

Delphinidin (blue/violet)

Betanidin (reddish)

Benzene is a nonpolar solvent. The many charges on betanidin make it likely

to be last. Of the other two, beta-carotene is so much larger that it is likely

to be second-to-last, even though delphinidin has some polar bonds.

What's So Special About Bottled Drinking Water?

Teacher Notes

TIME REQUIRED 45 min

SKILLS ACQUIRED

Collecting data
Communicating
Experimenting
Identifying patterns
Inferring
Interpreting
Organizing and analyzing data

RATING

Easy ⟵ 1   2   3   4 ⟶ Hard

Teacher Prep–2
Student Set-Up–2
Concept Level–2
Clean Up–2

THE SCIENTIFIC METHOD

Make Observations Students collect data indicating the amount of water absorbed by sodium polyacrylate for selected bottled drinking water samples.

Analyze the Results Analysis questions 1 and 2 and Conclusions question 1

Draw Conclusions Analysis question 1 and Conclusions questions 1 and 2

Communicate the Results Analysis questions 1 and 2

MATERIALS

In a common work area, set out bottled water from at least five separate sources. Include distilled water, mineral water, and imported and domestic drinking water.

   If a commercial source of sodium polyacrylate is unavailable, remove some from disposable diapers.

If materials are available, have students take multiple measurements for each sample and graph the average of the results.

SAFETY CAUTIONS

Read all safety precautions, and discuss them with your students.

• Safety goggles and a lab apron must be worn at all times.

• In case of a spill, use a dampened cloth or paper towels to mop up the spill. Then rinse the cloth or towels in running water at the sink, wring them out until only damp, and put them in the trash.

TIPS AND TRICKS

Pool class data for a more accurate graph. When recording marketing claims, students should pay careful attention to the use of nonregulated terminology such as *purity guaranteed* and *crisp, refreshing taste*. They should also note graphics that imply a specific source, such as a mountain stream or spring. You might point out to students that labels do not use the word *natural* as a standard of identity, because the FDA has not defined *natural*. Point out that seltzers, soda water, and tonic water are considered beverages, not bottled water. The amount of dissolved solid is reported either in parts per million (ppm) or in milligrams per liter (mg/L). These units are equivalent. Most imported brands and some domestic brands of bottled water are naturally carbonated because carbon dioxide gas present underground is dissolved in the water. Mineral salts help retain carbon dioxide. They bind with CO_2, resulting in water that has smaller bubbles. Most imported mineral water has very small gas bubbles. Domestic sparkling water also contains gas bubbles, but these may be larger and more numerous because of the recarbonation following purification.

Most brands do not identify a collection source on the label except for "bottled at the source." All domestic national, regional, and generic brands identify some type of purification; some brands list multiple types. Imported brands do not list any purification measures; they simply acknowledge that the source is protected. The European community requires mineral water to be free of specified chemical and microbiological impurities; it must be bottled at the source and cannot be filtered or treated in any way.

TECHNIQUES TO DEMONSTRATE

Demonstrate how to determine when to stop adding water to the polymer.

DISPOSAL

Water samples may be poured down the drain. Used polymer samples may be placed in the trash. Bottles should be placed in appropriate recycling containers.

Name _____ Class _____ Date _____

What's So Special About Bottled Drinking Water?

The International Bottled Water Association (IBWA) is the trade association that represents the bottled-water industry. Founded in 1958, IBWA's member companies produce and distribute more than 85% of the bottled water sold in the United States. Within the United States, bottled water is regulated as a food by the U.S. Food and Drug Administration (FDA). In contrast, municipal water is regulated as a commodity by the Environmental Protection Agency (EPA).

The FDA has labeling rules and regulations for bottled water. A table of definitions for types of bottled water appears on the next page.

Labels must include the name of the manufacturer, a statement of the net contents, and a list of ingredients if more than one ingredient is present. Any nutrient content claims must meet additional FDA regulations. *Sodium free* is an example of a nutrient content claim. When errors occur in the labeling of a product, or when a label intentionally misrepresents the product, the product has been misbranded. To correct a misbranding, the FDA works with the manufacturer through voluntary compliance, civil action, or criminal action, depending on the circumstances.

In this experiment, you will examine the labels of bottled-water samples and determine the amount of each sample that is absorbed by 0.1 g of sodium polyacrylate. The amount of water absorbed is related to the amount of mineral salts in the water.

OBJECTIVES

Conduct an audit of various brands of bottled water.

Compare the mineral content of various brands of bottled waters.

MATERIALS

- sodium polyacrylate
- balance
- 7 oz plastic cups
- 100 mL graduated cylinder
- microspoon or microspatula
- micropipet or medicine dropper
- wax pencil
- weighing boat
- bottled water, five types

Always wear safety goggles and a lab apron to protect your eyes and clothing. If you get a chemical in your eyes, immediately flush the chemical out at the eyewash station while calling to your teacher. Know the locations of the emergency lab shower and the eyewash station and the procedures for using them.

What's So Special About Bottled Drinking Water? *continued*

 Do not touch any chemicals. If you get a chemical on your skin or clothing, wash the chemical off at the sink while calling to your teacher. Make sure you carefully read the labels and follow the precautions on all containers of chemicals that you use. If there are no precautions stated on the label, ask your teacher what precautions you should follow. Do not taste any chemicals or items used in the laboratory. Never return leftovers to their original containers; take only small amounts to avoid wasting supplies.

 Call your teacher in the event of a spill. Spills should be cleaned up promptly, according to your teacher's directions.

Never put broken glass in a regular waste container. Broken glass should be disposed of properly.

Standardized Definitions for Types of Bottled Water

Artesian water/ artesian well water	bottled water from a well that taps a confined aquifer (a water-bearing underground layer of rock or sand) and allows water to flow naturally to the surface
Drinking water	bottled water sold for human consumption, containing no additives other than flavors, extracts, or essences less than 1% by mass; calorie free and sugar free
Mineral water	bottled water containing not less than 250 mg/L (ppm) total dissolved solids, with a constant level and regular proportion of mineral and trace element composition at the point of emergence from the source, and with no added minerals
Purified water	water produced by distillation, deionization, reverse osmosis, ozonation, or other suitable process; also called distilled water or reverse osmosis water
Spring water	bottled water derived from an underground formation that flows naturally to a spring or tap at the surface
Sparkling water	water that contains the same amount of carbon dioxide that it had at emergence from the source (does not pertain to seltzers, soda, or tonic water, which are soft drink beverages)
Well water	bottled water from a well, bored or drilled in the ground, that taps the water of an aquifer

What's So Special About Bottled Drinking Water? *continued*

Procedure
PART 1: CONDUCT A BOTTLED WATER AUDIT

1. Obtain five samples of bottled water from your teacher. Write the name of each sample in the **Data Table.**

2. Carefully read the labels on each sample product. Then record the data requested in the **Data Table.** Use the information in the Standardized Definitions for Types of Bottled Water table to identify the product label description.

PART 2: COMPARE THE MINERAL CONTENT OF THE BOTTLED WATER SAMPLES

3. Use a wax pencil to label a set of five transparent plastic cups *1* to *5*.

4. Add 0.1 g sodium polyacrylate to each cup. Tap the sides of the cup to spread the particles across the bottom of the cup.

5. Carefully open each sample bottle. Note whether the sample has gas bubbles. Record your findings in the **Data Table.**

6. Fill a graduated cylinder to the 100 mL mark with water from sample bottle *1.*

7. Using a micropipet or medicine dropper, remove water from the graduated cylinder and add it dropwise to the polymer in cup *1.* Keep adding drops until the white powder is no longer visible. *You may need to observe the underside of the cup carefully and swirl the beaker to ensure the absence of white powder.*

8. When the white powder is no longer visible, empty any water remaining in the micropipet or medicine dropper into the graduated cylinder. Record in the **Data Table** the volume of water absorbed by the white powder.

9. Repeat steps **6, 7,** and **8** for each of the remaining water samples.

10. On the grid provided, graph volume of water absorbed (mL) versus amount of dissolved solids (ppm) for each sample tested. Distilled water has 0.0 dissolved solids. Draw a line of best fit through the data points. Use this graph to estimate the amount of dissolved solids for any sample that does not indicate a value on its label.

DISPOSAL

11. Clean all apparatus and your lab station. Return equipment to its proper place. Dispose of chemicals and solutions in the containers designated by your teacher. Do not pour any chemicals down the drain or put them in the trash unless your teacher directs you to do so. Wash your hands thoroughly after all work is finished and before you leave the lab.

What's So Special About Bottled Drinking Water? *continued*

Data Table

	Sample 1 Aquafina	Sample 2 Distilled water	Sample 3 Gerolsteiner	Sample 4 Evian	Sample 5 Perrier
Product label description	purified drinking water	purified water	sparkling mineral water	spring water	sparkling mineral water
Carbonated	yes [ ] no [√]	yes [ ] no [√]	yes [√] no [ ]	yes [ ] no [√]	yes [√] no [ ]
Method of purification	reverse osmosis	distillation, ozonation, carbon filtration	none	none	none
Dissolved solids (mg/L or ppm)	not indicated	0.0	2527	309	475*
Product packaging	glass [ ] plastic [√] color [ ] no color [√]	glass [ ] plastic [√] color [ ] no color [√]	glass [√] plastic [ ] color [ ] no color [√]	glass [ ] plastic [√] color [ ] no color [√]	glass [√] plastic [ ] color [√] no color [ ]
Collection source on label	yes [ ] no [√]	yes [ ] no [√]	yes [√] no [ ]	yes [√] no [ ]	yes [√] no [ ]
Marketing claims	purity guaranteed; crisp refreshing taste	none	low sodium; contains natural calcium; naturally sparkling	as pure and natural as when its source was discovered	low mineral content
Size of carbonation bubbles	N/A	N/A	tiny	N/A	small
Volume of water absorbed (mL)	52	65	3	42	25

* from www.perrier.com

What's So Special About Bottled Drinking Water? *continued*

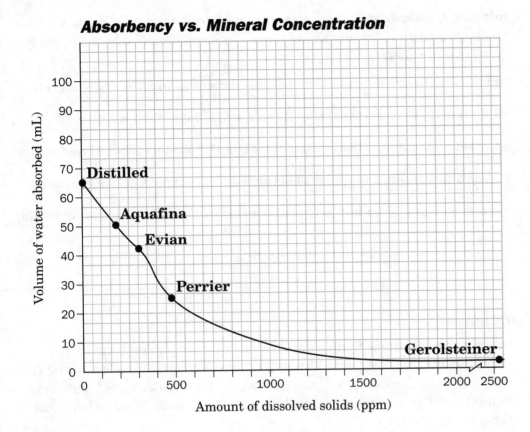

Absorbency vs. Mineral Concentration

Analysis

1. Analyzing Data How is the volume of water absorbed, known as absorbency, related to the amount of dissolved solids?

Absorbency decreases as the dissolved solid concentration (mineral content) increases. Aquafina's label did not indicate its dissolved solids. It had an absorbency of approximately 52 mL. Its dissolved solids content is approximately 200 mg/L.

What's So Special About Bottled Drinking Water? *continued*

2. **Inferring Conclusions** If your data were used to determine compliance with labeling regulations, would any of the water samples tested be misbranded? Justify your answer.

Answers will vary. Examples of misbranding are as follows: If high levels of

mineral salts were present, more than a single ingredient would be present,

and there should be a list of ingredients on the product label. If no list were

present, the product would be misbranded. If a sample were labeled as

sodium free but had a low polymer absorbency level, further testing would

be needed.

Conclusions

1. **Inferring Conclusions** Spring water tends to be the predominant type of domestic and regional bottled water. Very few domestic brands are labeled *mineral water;* most bottled mineral water is imported. Suggest a reason why domestic companies do not market mineral water even though FDA regulations permit them to.

Mineral water lacks market appeal. Apparently, Americans prefer spring

water, perhaps because the term sounds fresh and pure. Accept all reason-

able answers citing economic or marketing issues.

2. **Analyzing Data and Inferring Conclusions** Examine your packaging data. Which type of water comes in colored containers? Suggest a reason for the color. (Hint: Examine the method-of-purification data.)

Mineral water comes in colored containers. The color protects the water

from sunlight, which causes algal growth.

How Effective Are Antacids?

Teacher Notes

TIME REQUIRED One 45-minute lab period

SKILLS ACQUIRED
Collecting data
Experimenting
Identifying patterns
Inferring
Interpreting
Measuring
Organizing and analyzing data

RATING

Easy ◄——1——2——3——4——► Hard

Teacher Prep–3
Student Set-Up–3
Concept Level–3
Clean Up–2

THE SCIENTIFIC METHOD

Make Observations Students collect titration and pH data

Analyze the Results Analysis questions 1 and 2

Draw Conclusions Conclusions question 1 and Analysis question 2

Communicate the Results Analysis question 2 and Conclusion question 2

MATERIALS

Materials for this lab activity can be purchased from Sargent-Welch. See the Lab Materials QuickList software on the **One-Stop Planner** CD-ROM for ordering instructions.

To prepare 1 L of 0.5 M HCl solution, observe the required safety precautions. While stirring, slowly add 41.5 mL of 12 M HCl to 500 mL of distilled water. Dilute to 1 L.

To prepare 1 L of 1 M NaOH solution, observe the required safety precautions. While stirring, slowly add 40 g of NaOH to 800 mL of distilled water. Stir to dissolve the solid. Once dissolved, add distilled water to a final volume of 1 L.

Select 5 antacid brands; include both liquids and tablets. Set containers in a general access area. Use one tablet or 1 tsp (5 mL) per sample.

SAFETY CAUTIONS

Wear safety safety gogles, a face shield, impermeable gloves, and an apron when you prepare the HCl and NaOH solutions. Work in a chemical fume hood known to be in operating condition, and have another person stand by to call for help in case of an emergency. Be sure you are within a 30 s walk from a safety shower and eyewash station known to be in good operating condition.

Remind students of the following safety precautions:

• Always wear safety goggles and a lab apron to protect your eyes and clothing. If you get chemical in your eyes, immediately flush the chemical out at the eye-wash station while calling to your teacher. Know the location of the emergency lab shower and the eyewash stations and procedures for using them.

• Do not touch any chemicals. If you get a chemical on your skin or clothing, wash the chemical off at the sink while calling to your teacher. Make sure you carefully read the labels and follow the precautions on all containers of chemicals that you use. If there are no precautions stated on the label, ask your teacher what precautions you should follow. Do not taste any chemicals or items used in the laboratory. Never return leftovers to their original containers; take only small amounts to avoid wasting supplies.

• Call your teacher in the event of a spill. Spills should be cleaned up promptly, according to your teacher's directions. In case of a spill, use a dampened cloth or paper towels to mop up the spill. Then rinse the cloth in running water at the sink, wring it out until it is only damp, and put it in the trash.

• Never put broken glass in a regular waste container. Broken glass should be disposed of properly.

DISPOSAL

Set out three disposal containers for the students: one for unused acid solutions, one for unused base solutions, and one for partially neutralized substances and the contents of the waste beaker. One at a time, slowly combine solutions while stirring. Adjust the pH of the final waste liquid with 1 M acid or base until the pH is between 5 and 9. Pour the neutralized liquid down the drain.

TECHNIQUES TO DEMONSTRATE

• Show the students the proper method for filling a buret. Strongly caution students against reaching above their head to fill a buret.

• Demonstrate the end point of bromthymol blue or the use of a pH meter.

• Review how to read a meniscus and how to clean a buret.

TIPS AND TRICKS

Computing the milliequivalents of acid neutralized by antacids

Discuss the back-titration process. Students should understand why HCl is added in excess and why NaOH is the titrant. They should also understand how to determine the end point of a titration. Students also should understand why the acid/antacid mixtures are allowed to stand for 15 minutes.

Analysis

Discuss the difference between free acid and combined acid. Students should be aware that they are titrating free acid rather than combined acid.

How Effective Are Antacids?

An acidic stomach is necessary for good health, but excessive stomach acid can produce acute discomfort or contribute to ulcers. An effective antacid neutralizes just enough excess acid to alleviate pain and discomfort; it does not bring stomach acids to neutrality (pH 7.0).

The bases most widely used as active ingredients in antacids are of two types: absorbable and nonabsorbable. Absorbable antacids include $NaHCO_3$ and $CaCO_3$ and are very effective at increasing gastric pH. Because they are easily absorbed into the blood, these compounds can also raise blood pH to dangerous levels, causing kidney damage. Nonabsorbable antacids are relatively insoluble salts of weak bases, such as $Al(OH)_3$ and $Mg(OH)_2$. They interact with HCl, forming nonabsorbed or poorly absorbed salts while increasing gastric pH.

Acid secreation is frequently divided into free acid and combined acid. The amount of free acid is determined by titrating gastric secretions to a pH of 3.5. After this titration is performed, the same gastric secretions are titrated to a pH of 8.5; this measures the combined acid. Gastric secretions mixed with food show little or no free acid but a large amount of combined acid. Conversely, when the stomach secretes large quantities of gastric juice while it is almost empty of food, the larger portion of acid is free acid. In this experiment, you will add HCl to antacid samples and then neutralize the excess HCl by titrating to a pH of 3.5 with NaOH.

OBJECTIVES

Compare the neutralization ability of antacids.

Infer which antacid(s) tested is most effective.

MATERIALS

- antacids, tablets and liquid
- beaker, 150 mL (5)
- beaker, 250 mL (2)
- buret clamp
- buret, 50 mL
- graduated cylinder, 10 mL
- HCl, 0.5 M
- microspatula

- mortar and pestle
- NaOH, 1 M
- pH meter or bromthymol blue indicator
- ring stand
- stirring rod
- wax pencil

Always wear safety goggles, gloves, and a lab apron to protect your eyes and clothing. If you get a chemical in your eyes, immediately flush the chemical out at the eyewash station while calling to your teacher. Know the location of the emergency lab shower and eyewash station and the procedures for using them.

| How Effective Are Antacids? *continued*

 Do not touch any chemicals. If you get a chemical on your skin or clothing, wash the chemical off at the sink while calling to your teacher. Make sure you carefully read the labels and follow the precautions on all containers of chemicals that you use. If there are no precautions stated on the label, ask your teacher what precautions to follow. Do not taste any chemicals or items used in the laboratory. Never return leftovers to their original container; take only small amounts to avoid wasting supplies.

 Call your teacher in the event of a spill. Spills should be cleaned up promptly, according to your teacher's directions.

Acids and bases are corrosive. If an acid or base spills onto your skin or clothing, wash the area immediately with running water. Call your teacher in the event of an acid spill. Acid or base spills should be cleaned up promptly.

Never put broken glass in a regular waste container. Broken glass should be disposed of separately according to your teacher's instructions.

Procedure

PART 1—COMPUTING AMOUNT OF ACID NEUTRALIZED

1. Put on safety goggles, gloves, and a lab apron.

2. Use a wax pencil to label a set of five 150 mL beakers 1, 2, 3, 4, and 5.

3. Obtain five antacid samples from your teacher. Record the manufacturer's brand name and the active ingredients in **Table 1**. Then prepare one sample per beaker.

   For antacid tablets: Crush one tablet, using a mortar and pestle. Using a microspatula, transfer the contents to the appropriately numbered beaker. To the crushed tablet, add 100 mL of 0.5 M HCl. Use a stirring rod to help dissolve the crushed tablet.

   For liquid antacids: Pour 5 mL of the antacid into a graduated cylinder. Add 0.5 M HCl to bring the final volume to 100 mL. Then pour this mixture into the appropriately labeled beaker. Use a stirring rod to mix. Allow the mixtures to stand for 15 minutes. Stir occasionally.

4. While the mixtures are standing, prepare the titration apparatus. Attach a buret clamp to a ring stand. Use a wax pencil to label the buret 1 M NaOH. Insert the buret in the buret clamp.

5. Label a 250 mL beaker "Waste." Fill a second 250 mL beaker with approximately 125 mL of 1 M NaOH. Carefully pour 5 mL of the 1 M NaOH from the beaker into the buret. Rinse the walls of the buret thoroughly with this solution. Allow the solution to drain through the stopcock into the waste beaker. Close the stopcock. Rinse the buret two more times in this manner, using a new 5 mL portion of NaOH solution each time.

How Effective Are Antacids? *continued*

6. Fill the buret above the zero mark with 1 M NaOH. Place the waste beaker under the buret, and withdraw enough solution to remove any air from the buret tip and to bring the liquid level within the graduated region of the buret. Record the initial volume of NaOH in **Table 1**.

7. If you are using a pH meter, calibrate the pH probe according to your teacher's instructions. Then insert the pH probe into the beaker containing the first antacid sample to be analyzed. If you are using bromthymol blue indicator, add 10 drops of the indicator to the acid/antacid mixture. Gently swirl the beaker to mix.

8. Place the beaker from step 7 under the buret, and add NaOH dropwise until a pH reading of 3.5 is reached. Record the final buret reading in **Table 1**. If using a pH meter, remove the pH probe. Rinse it with distilled water before placing it in the next sample.

9. Repeat steps 6 through 8 for the remaining antacid samples.

10. Clean all apparatus and your lab station. Return equipment to its proper place. Dispose of chemicals and solutions in the containers designated by your teacher. Do not pour any chemicals down the drain or put them in thetrash unless your teacher directs you to do so. Wash your hands thoroughly after all work is finished and before you leave the lab.

Analysis

1. **Organizing Data** The FDA requires manufacturers to rate the effectiveness of their antacid in terms of the amount of acid neutralized. The amount of acid neutralized is expressed in milliequivalents (mEq). Calculate the mEq of acid neutralized by each antacid using the following equation:

mEq of acid neutralized = $[V_{HCl\ used} \times C_{HCl}] - [V_{NaOH\ added} \times C_{NaOH}]$

If 10 mL of base is added, then the mEq of acid neutralized is:

[(30 mL HCl) × (0.5 M HCl)] − [(10 mL NaOH) × (1 M NaOH)] =

15 − 10 = 5 mEq of acid neutralized.

How Effective Are Antacids? *continued*

TABLE 1: COMPARING THE NEUTRALIZATION ABILITY OF ANTACIDS

Sample number	Buret readings		Volume of NaOH used (mL NaOH)	mEq of acid neutralized	Antacid brand bame	Antacid active ingredients
	Initial	Final				
1	2.0	36.0	34.0	16	Tums Extra Strength tablet	750 mg CaCO$_3$
2	0.0	31.1	31.1	18.9	Maalox Extra Strength tablet	1000 mg CaCO$_3$; 60 mg simethicone
3	0.0	33.5	33.5	16.5	Rolaids tablet	550 mg CaCO$_3$; 110 mg Mg(OH)$_2$
4	0.0	33.0	33.0	17	Equate	750 mg CaCO$_3$
5	0.0	39.4	39.4	10.6	Diovol Plus AF liquid	200 mg CaCO$_3$; 200 mg Mg(OH)$_2$; 25 mg simethicone

2. **Constructing Graphs** Use the grid in **Figure 1** below to draw a bar graph that summarizes the milliequivalents of acid neutralized by brand sample. Along the horizontal axis, indicate the brand name. Along the vertical axis, record the milliequivalents of acid neutralized either per tablet or per 5 mL of liquid.

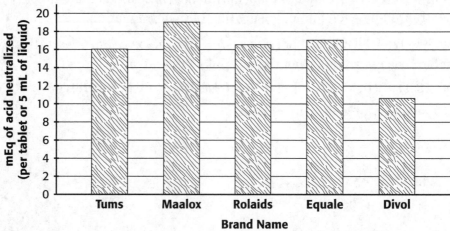

Figure 1

3. **Applying Data** Based on the graph summarizing the milliequivalents of acid neutralized by an antacid tablet or liquid dose, which brand is the most effective? Support your answer.

Answers will vary depending on the brands tested. The brand with the

largest milliequivalents per tablet or 5 mL dose should be identified.

How Effective Are Antacids? *continued*

Conclusions

1. Drawing Conclusions Using data from the Data Table are the liquid forms of popular antacids more effective than tablets in neutralizing acid?

Generally, the data suggest that this relationship does exist.

2. Applying Conclusions Copy the advertising claims made on the antacid brands evaluated. Are the claims of "extra strength" supported by your data? Does this experiment allow you to evaluate claims for "quicker acting"? Support your answers.

Claims of extra strength should be supported by data because these products

contain larger amounts of active ingredients. No, time to dissolve or react

were not examined, so the data cannot be used to evaluate claims of quick-

acting formulas.

Shampoo Chemistry

Teacher Notes

TIME REQUIRED 45 min

SKILLS ACQUIRED
Classifying
Collecting data
Communicating
Experimenting
Identifying patterns
Inferring
Interpreting
Organizing and analyzing data

RATING

Easy ◄——1——2——3——4——► Hard

Teacher Prep–2
Student Set-Up–3
Concept Level–2
Clean Up–2

THE SCIENTIFIC METHOD

Make Observations Students observe the properties of various brands of shampoo.

Analyze the Results Analysis questions 1 and 2

Draw Conclusions Conclusions questions 1 to 3

Communicate the Results Analysis questions 1 and 2

MATERIALS

Set out containers of shampoo.

SAFETY CAUTIONS

Read all safety precautions, and discuss them with your students.

• Safety goggles and a lab apron must be worn at all times.

Shampoo Chemistry *continued*

TIPS AND TRICKS

Obtain various clear shampoo brands for student analysis. Include baby shampoos and shampoos that are also conditioners. Use multiple product containers so students can easily review ingredient listings.

If time or shampoo supplies are short, have each student study three shampoo samples instead of five.

Discuss the importance of foam, and review calculation of the foam-to-liquid ratio. Make sure students understand why the volume of foam is divided by 25 mL.

Remind students to keep a thumb over the stopper when shaking the graduated cylinder.

Remind students that FDA labeling regulations require that the highest concentration ingredient be listed first, followed by the next highest, and so on.

DISPOSAL

Recover the BBs, rinse them thoroughly, dry them, and store them in an appropriately labeled container for later use. Unused shampoos and India ink should be stored for use at a later time. Shampoo solutions can be washed down the drain with lots of water.

Inquiry

Shampoo Chemistry

Hair is a lifeless structure composed of the cross-linked polymer protein keratin. Cross-linking gives hair its strength. A strand of hair has a central core (cortex) that contains its coloring pigments. A thin, scaly sheath, called the cuticle, encases the cortex. Within the scalp, sebaceous glands secrete an oily substance (sebum), which gives hair its gloss and keeps the scales of the cuticle lying flat. Too much sebum makes the hair feel greasy and dirty. Too little sebum makes the hair dry and unmanageable.

The principle function of a shampoo is to produce lather and remove dirt from the hair and scalp without removing all of the oils. Lather is the foam created when a surfactant is agitated in water. A good shampoo will have at least twice its volume in lather, or a 2:1 ratio. Lather stabilizers help maintain the foam condition during product use. The longer the foam lasts, the more effective the cleansing action is. Dirt and grease trapped in the foam fraction of the shampoo are more difficult to rinse away than dirt and grease dispersed in the liquid fraction. India ink contains a dispersion of minute carbon particles in water. In this experiment, you will use India ink as a test dirt. By observing whether the tiny carbon particles are trapped in the shampoo's foam, you will evaluate the cleansing effectiveness of the product.

We also expect our shampoos to provide fullness and luster to our hair. The pH of a shampoo is responsible for luster. Acidic pH shampoos tend to tighten the hair cuticle and scales, allowing light to reflect evenly. Basic pH shampoos tend to swell cuticle scales, resulting in light that leaves hair looking "dull" and "flat." Shampoos with pH values between 4.5 and 6.5 produce the best luster, or shiniest hair. The body of a shampoo is its viscosity. The higher the viscosity is, the thicker the shampoo—a quality that implies a premium product.

OBJECTIVES

Compare selected shampoos for the following characteristics: pH, foam volume and retention, oil dispersal, and viscosity.

Identify shampoo ingredients.

MATERIALS

- 150 mL beakers, 5
- 100 mL graduated cylinder
- clear shampoo samples
- India ink
- copper BBs or small steel ball bearings
- medicine dropper
- No. 3 stopper
- stopwatch or watch with a second hand
- universal pH paper
- wax pencil

| Shampoo Chemistry *continued*

  Always wear safety goggles and a lab apron to protect your eyes and clothing. If you get a chemical in your eyes, immediately flush the chemical out at the eyewash station while calling to your teacher. Know the locations of the emergency lab shower and the eyewash station and the procedures for using them.

Do not touch any chemicals. If you get a chemical on your skin or clothing, wash the chemical off at the sink while calling to your teacher. Make sure you carefully read the labels and follow the precautions on all containers of chemicals that you use. If there are no precautions stated on the label, ask your teacher what precautions you should follow. Do not taste any chemicals or items used in the laboratory. Never return leftovers to their original containers; take only small amounts to avoid wasting supplies.

Call your teacher in the event of a spill. Spills should be cleaned up promptly, according to your teacher's directions..

Never put broken glass in a regular waste container. Broken glass should be disposed of properly.

Procedure
PART 1: COMPARING SHAMPOO CHARACTERISTICS

1. **Preparing a 1% Shampoo Solution** Obtain from your teacher five shampoo samples. Use a wax pencil to label five beakers *1* to *5*. Add 50.0 mL of deionized water to a 100 mL graduated cylinder, then add 2.0 mL of a shampoo.

2. Fill the cylinder to the 100 mL mark with deionized water. Pour the diluted shampoo into a numbered beaker. Thoroughly rinse the graduated cylinder between shampoo samples.

3. **Testing pH** Using universal pH test paper, determine the pH value of each shampoo sample. Record the values in **Data Table 1.** Which shampoo brand(s) would give hair the best shine, or luster?

   Answers will vary. Chosen brand(s) should have pH values that are slightly

   acidic. Basic pH shampoos tend to dull hair.

4. **Testing Lather Volume and Retention** Carefully pour 25 mL of one diluted shampoo sample into a 100 mL graduated cylinder. To avoid producing foam, slowly pour the diluted shampoo sample down the *side* of the graduated cylinder.

5. Firmly stopper the cylinder. Shake the cylinder up and down 10 times. Record the volume of foam in **Data Table 1.** Calculate the ratio of foam to liquid by dividing the volume of foam by 25 mL. Enter this ratio in **Data Table 1.**

6. Remove the stopper from the graduated cylinder. Record in **Data Table 1** the volume of foam remaining after 1 min and after 5 min.

Shampoo Chemistry *continued*

7. Repeat steps **4** through **6** for the remaining diluted shampoo samples. Do all shampoos pass the 2:1 ratio test for lather? Justify your answer.

   Answers will vary. Most shampoos should maintain a 2:1 ratio for at least 1

   min. Check that the data support the justification.

8. **Testing Dispersal** Carefully pour 25 mL of one diluted shampoo sample into a 100 mL graduated cylinder. Avoid foaming. Add one drop of India ink, stopper the cylinder, and shake the cylinder up and down 10 times. Remove the stopper from the graduated cylinder. Determine if the India ink has been dispersed (evenly spread) throughout either the liquid portion or the foam portion of the shampoo sample. Place a check mark next to "liquid" and/or "lather" in **Data Table 1.**

9. **Testing Shampoo Viscosity** Fill a graduated cylinder to the 100 mL mark with an *undiluted* shampoo sample. Release a copper BB or small steel ball at the top of the cylinder. Using a stopwatch or a watch with a second hand, determine the time for the BB to fall to the bottom of the cylinder. The longer the drop time is, the thicker and more viscous the shampoo. Record the time in **Data Table 1.**

10. Repeat steps **8** and **9** for the other undiluted shampoo samples.

Data Table 1

Sample number and brand	pH	Foam volume (mL) and foam ratio	Foam retention/ volume of foam (mL)		Ink dispersal (√)	Viscosity [BB drop time] (s)
1. Jergens Moisturizing Body Shampoo	8.0	volume: 80 ratio: 3.2:1	1 min 80	5 min 75	liquid [√] lather [ ]	3
2. Alberto V05	6.5	volume: 80 ratio: 3.2:1	1 min 80	5 min 78	liquid [√] lather [ ]	2.5
3. St. Ives Swiss Formula	7.5	volume: 83 ratio: 3.3:1	1 min 82	5 min 79	liquid [√] lather [ ]	23
4. Revlon Flex Balsam and Protein	7.5	volume: 90 ratio: 3.6:1	1 min 90	5 min 90	liquid [√] lather [ ]	2
5. Thermasilk	6.5	volume: 87 ratio: 3.5:1	1 min 86	5 min 84	liquid [√] lather [ ]	3

Shampoo Chemistry *continued*

PART 2: PERFORMING AN "INGREDIENT" AUDIT

11. Examine the ingredients listing for at least one shampoo product. Using the **Information Table** as a guide, determine in which functional category each ingredient belongs, and write the name of the ingredient in the matching column of **Data Table 2.** If you have an ingredient whose function is not found below, use an internet search engine to determine its function.

Information Table

Ingredient	Weight %	Functional category and function
Purified water • aqua	60	solvent
Triethanolammonium lauryl sulfate • sodium lauryl sulfate • sodium laureth sulfate • ammonium lauryl sulfate • ammonium laureth sulfate • cocamidopropyl betaine • cococamphodiacetate • sodium cocglyceryl ether sulfonate • sodium lauryl sarcosinate • sodium laureth–13	32	surfactants (clean hair)
Myristic acid • citric acid	4	pH adjusters (increase luster)
Cetyl alcohol • stearyl alcohol • caprylic acid • Hydrogenated lanolin • polyethylene glycol • glycol sterate • palmitic acid • PEG-80		thickeners (give texture, appearance, and flow to the product)
Oleyl alcohol • panthenol • amino acids • protein • collagen • dimethicone	2	conditioning agents (give hair a softer, thicker feel)
Cocamide MEA • lauramide MEA • lauric DEA • polysorbate-20	1	lather stabilizers (maintain lather during use)
Fragrance (parfum)	1	perfumes
Guar hydroxypropyltrimonium chloride • dicetydimonium chloride • behentrimonium chloride • behanalkonium betaine • benzalkonium chloride • quaterium-18 • stearalkonium chloride •cetrimonium chloride • polyquaterium 10 • quaterium 15	1	quaternary ammonium compounds (detangle hair)
Formaldehyde • methyl paraben • propylparaben • phenoxyethanol • quateternium 15 • DMDM hydantoin	0.5	preservatives (act as antimicrobials)
Glycerin • sorbitol • glycols • mucopolysaccharides • hyaluronic acid • sodium PCA • glycoshingolipids • sorbitan laurate	0.5	humectants (hold water in the hair shaft, giving hair bounce and fullness)
Polyphosphates • EDTA	0.5	sequestrants and chelating agents (soften hard water)
FD&C blue, red, yellow	0.5	colorants

Shampoo Chemistry *continued*

Data Table 2

Product Name: Revlon Flex Balsam and Protein

Functional category	Present in ingredients listing (√)	Named ingredients
Solvent	√	water
Surfactant(s)	√	ammonium lauryl sulfate, sodium C14-17 alkyl SEC sulfonate, sodium laureth sulfate, cocamidepropyl betaine
pH adjustment	√	citric acid
Thickeners	√	methyl cellulose, hydroxyl propyl methyl cellulose
Conditioning agents	√	panthenol
Lather stabilizers	√	lauramide DEA
Perfumes	√	fragrance
Preservatives	√	methyl paraben, propyl paraben
Humectants		
Sequestrant(s) and chelating agents	√	tetrasodium EDTA
Colorants	√	FD&C yellow #5, D&C red #33, FD&C red #4, FD&C yellow #6

DISPOSAL

Clean all apparatus and your lab station. Return equipment to its proper place. Dispose of chemicals and solutions in the containers designated by your teacher. Do not pour any chemicals down the drain or put them in the trash unless your teacher directs you to do so. Wash your hands thoroughly after all work is finished and before you leave the lab.

Shampoo Chemistry *continued*

Analysis

1. Analyzing Data According to entries in the **Information Table,** which ingredients in a shampoo play no part in the removal of oils and dirt?

pH adjusters, conditioning agents, perfumes, quaternary ammonium com-

pounds, preservatives, and colorants play no part in dirt and oil removal.

Surfactants are the primary cleansing agents, supported by lather stabilizers.

Sequestrants and chelating agents allow the surfactants to work in hard-

water conditions.

2. Analyzing Data Which shampoo brand(s) have the greatest viscosity? Do these same brands bear advertising that promotes extra body?

Answers will vary. BB drop time should be about equal except for brands

that advertise extra body.

Conclusions

1. Inferring Conclusions What do you think is meant by the advertising term *natural pH*?

a pH that is slightly acidic

2. Analyzing Information Which shampoo brand provides the best value? Give reasons for your answer.

Answers will vary. Justifications may include inexpensiveness, pH that

promotes luster, least amount of dirt in foam, or best foaming action. Accept

all reasonable answers.

3. Inferring Conclusions Examine the **Information Table** entries and compare them to the ingredients on a bottle of conditioner. Suggest reasons for differences in the composition of shampoos and conditioners.

Conditioners will contain a larger number of thickeners, humectants, and

conditioning agents than shampoos; conversely, conditioners tend to have

fewer surfactants.

Solubility Product Constant

Teacher Notes

TIME REQUIRED One 45-minute lab period

SKILLS ACQUIRED
Collecting data
Communicating
Experimenting
Identifying patterns
Inferring
Interpreting
Organizing and analyzing data

RATING
Easy ⟵——1——2——3——4——⟶ Hard

Teacher Prep–3
Student Set-Up–2
Concept Level–3
Clean Up–2

THE SCIENTIFIC METHOD

Make Observations Students collect solubility data.

Analyze the Results Analysis questions 1 to 8 and Conclusions questions 9 and 10

Draw Conclusions Conclusions questions 9 and 10

Communicate the Results Conclusions questions 9 and 10

MATERIALS

Sodium citrate is actually used as a food additive in some products. However, if it is not in stock, the common-ion portion of the lab (**steps 10 and 11**) can be performed with the following other sodium salts instead of sodium citrate: sodium carbonate, sodium hydrogen carbonate, sodium bromide, sodium iodide, sodium oxalate, sodium sulfate, sodium thiosulfate, and sodium acetate. Tell the students which one will be used so that they can adjust their calculations accordingly. **Do not use sodium nitrate or other sodium salts besides those listed because of the difficulties involved in disposal.**

Instead of decanting in **step 3,** you may have students use micropipets to transfer an appropriate amount of the solution to the next beaker for testing. Be certain students do not place the tip of the pipet near the solid residue, or some of the solid will be drawn into the pipet and the results will not be as accurate.

SAFETY CAUTIONS

Safety goggles and a lab apron must be worn at all times.

Tie back long hair and loose clothing when working in the lab.

Read all safety cautions, and discuss them with your students.

Remind students of the following safety precautions:

- Always wear safety goggles, gloves, and a lab apron to protect your eyes and clothing. If you get a chemical in your eyes, immediately flush the chemical out at the eyewash station while calling to your teacher. Know the location of the emergency lab shower and the eyewash stations and procedures for using them.

- Do not touch any chemicals. If you get a chemical on your skin or clothing, wash the chemical off at the sink while calling to your teacher. Make sure you carefully read the labels, and follow the precautions on all containers of chemicals that you use. If there are no precautions stated on the label, ask your teacher what precautions you should follow. Do not taste any chemicals or items used in the laboratory. Never return leftover chemicals to their original containers; take only small amounts to avoid wasting supplies.

- When using a Bunsen burner, confine long hair and loose clothing. Do not heat glassware that is broken, chipped, or cracked. Use tongs to handle heated glassware and other equipment.

- Call your teacher in the event of a spill. Spills should be cleaned up promptly, according to your teacher's directions.

- Never put broken glass in a regular waste container. Broken glass should be disposed of properly.

DISPOSAL

Set out one disposal container for students. It should contain only sodium chloride and another sodium salt. If the other salt is one recommended in the Materials section, the salts can be dissolved in water and poured down the drain.

TECHNIQUES TO DEMONSTRATE

Have the equipment set up for heating the evaporating dish. If you are using a Bunsen burner, show students the appropriate size and type of flame needed to heat the salt solution slowly and avoid spattering. If you are using a hot plate, suggest a setting that will heat slowly without spattering.

Demonstrate the decanting step, and explain how important it is to avoid getting crystals of NaCl in the saturated solution that will be tested. If you will be using micropipets instead of decanting, point out how important it is to keep the tip of the pipet well above the solid so that none is accidentally included.

TIPS AND TRICKS

Be certain students understand that when they decant the saturated solution in **step 3,** it is the solution they want to keep, not any solid that remains undissolved. Students sometimes become confused about what they are supposed to do in this step.

Discuss the nature of solubility equilibrium in terms of the undissolved solid on the bottom of a beaker dissolving into the solution at the same rate that the dissolved solute precipitates out of solution. Students must understand that the apparently static nature of the system on the macroscopic level does not reflect the dynamic equilibrium at the microscopic level.

Solubility Product Constant

SITUATION

Juliette Brand Foods is planning to add dill pickles to their product line. Pickled foods require soaking or packing in a concentrated solution of sodium chloride called brine. The company intends to make their own brine in large quantities. To help them set up their process, they have asked your company to determine the mass of NaCl that should be added to water in a 4000.0 L vat at room temperature so that the solution will have the highest possible concentration with no NaCl remaining undissolved. Juliette Brand Foods is also considering the use of sodium citrate, $Na_3C_6H_5O_7$, as a flavoring agent. They need you to determine what effects, if any, on the solubility of NaCl the addition of $Na_3C_6H_5O_7$ will have.

BACKGROUND

A saturated solution contains the maximum amount of solute that can dissolve when added to a solvent at a given temperature. Saturated solutions usually contain an excess of undissolved solid so that the ions in solution are at equilibrium with the undissolved solid. The equilibrium constant that describes this situation is called the solubility product constant, K_{sp}. For salts composed of ions with single positive or negative charges, the K_{sp} is the product of the concentrations of dissolved ions. If either of the ion concentrations is increased after a saturated solution is formed, the value of the K_{sp} will be exceeded, and precipitation will occur until equilibrium is restored. This phenomenon is called the "common-ion effect."

PROBLEM

To answer your client's questions you need to do the following.

- Make a saturated solution of NaCl at room temperature.
- Calculate the concentration in moles per liter of each ion in the solution.
- Write the equilibrium expression and calculate the K_{sp} of NaCl at room temperature.
- Observe the effect of adding $Na_3C_6H_5O_7$ to a saturated solution of NaCl.
- Use your data to calculate the mass of NaCl necessary to fill the pickling vat with 4000.0 L of saturated solution.

OBJECTIVES

Prepare a saturated solution of sodium chloride.

Determine the mass of sodium chloride dissolved in a volume of saturated solution.

Calculate the solubility of sodium chloride.

Determine the solubility product constant of sodium chloride.

Demonstrate the effect of the addition of a common ion to a saturated solution.

❚ Solubility Product Constant *continued*

MATERIALS

- balance
- beakers, 150 mL (2)
- crucible tongs
- distilled water
- evaporating dish
- glass stirring rod
- graduated cylinder, 10 mL
- $Na_3C_6H_5O_7 \cdot 2H_2O$
- NaCl
- spatula

- test tube, 25 mm × 100 mm
- weighing paper

Bunsen Burner option

- Bunsen burner and related equipment
- ring stand and ring
- wire gauze with ceramic center (2)

Hot Plate option

- hot plate

Always wear safety goggles, gloves, and a lab apron to protect your eyes and clothing. If you get a chemical in your eyes, immediately flush the chemical out at the eyewash station while calling to your teacher. Know the location of the emergency lab shower and eyewash station and the procedures for using them.

Do not touch any chemicals. If you get a chemical on your skin or clothing, wash the chemical off at the sink while calling to your teacher. Make sure you carefully read the labels and follow the precautions on all containers of chemicals that you use. If there are no precautions stated on the label, ask your teacher what precautions to follow. Do not taste any chemicals or items used in the laboratory. Never return leftover chemicals to their original containers; take only small amounts to avoid wasting supplies.

Do not heat glassware that is broken, chipped, or cracked. Use tongs or a hot mitt to handle heated glassware and other equipment because hot glassware does not always look hot.

When using a Bunsen burner, confine long hair and loose clothing. If your clothing catches on fire, WALK to the emergency lab shower and use it to put out the fire.

Procedure

1. Put on safety goggles, gloves, and a lab apron.

2. Make sure that your equipment and tongs are very clean for this work so that you will get the best possible results. Remember that you will need to cool the evaporating dish before measuring its mass. **Never put a hot evaporating dish on a balance; it may damage the balance, and the currents of hot air will cause an error in the mass measurement.**

| **Solubility Product Constant** *continued*

3. Prepare a saturated solution of NaCl by adding 15 g of the salt to 25 mL of water in a beaker. Stir constantly until it appears that no more salt will dissolve. There should be some solid on the bottom of the beaker. Decant the saturated solution into another beaker, making certain that no solid is transferred. Pour the excess solid and the remaining solution down the drain with plenty of water.

4. Using tongs, place the evaporating dish on the balance and measure its mass to the nearest 0.01 g. Record the mass in **Table 1.**

5. Using a graduated cylinder, measure approximately 10.0 mL of the salt solution. Record the volume to the nearest 0.1 mL in **Table 1.** Pour this sample of saturated NaCl solution into the evaporating dish.

6. Set up the ring stand assembly and Bunsen burner as shown in **Figure 1.** The ring and wire gauze should be positioned several centimeters above the burner flame. Place the evaporating dish on the wire gauze, and light the burner. Heat the NaCl solution slowly to avoid spattering. Continue heating until all of the water has evaporated and the remaining crystals are dry.

7. When the remaining crystals appear dry, adjust the burner to produce a very low flame and continue heating for another 10 min. Hold the burner by the base, and gently wave the flame under the evaporating dish, being careful not to heat the material so rapidly that it spatters.

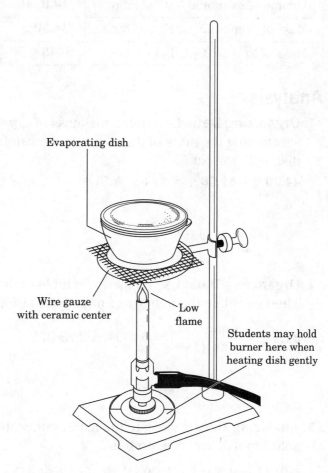

Evaporating dish

Wire gauze
with ceramic center

Low
flame

Students may hold
burner here when
heating dish gently

Figure 1

8. Using tongs, place the hot evaporating dish on a piece of wire gauze at the base of the ring stand and allow it to cool. Then determine the mass of the evaporating dish and the solid residue to the nearest 0.01 g. Record the mass in **Table 1.**

9. Measure 5.0 mL of saturated NaCl solution, and pour it into a test tube.

10. Using a spatula, measure about 0.50 g $Na_3C_6H_5O_7 \cdot 2H_2O$ on the balance. Record the exact mass in your data table.

11. Add the $Na_3C_6H_5O_7 \cdot 2H_2O$ to the test tube and stir. Record your observations.

Observations:

white precipitate forms

12. All of the solutions in this lab may be disposed of by washing them down the drain. Place any excess NaCl or $Na_3C_6H_5O_7 \cdot 2H_2O$ in the disposal container designated by your teacher.

TABLE 1 MASS DATA

Mass of empty evaporating dish	**41.36 g**
Volume of saturated NaCl solution	**10.0 mL**
Mass of evaporating dish and NaCl	**44.50 g**
Mass of $Na_3C_6H_5O_7 \cdot 2H_2O$	**0.49 g**

Analysis

1. Organizing Data Determine the mass of dry NaCl in the evaporating dish by subtracting the mass of the evaporating dish from the mass of the evaporating dish with the NaCl.

44.50 g − 41.36 g = 3.14 g NaCl

2. Organizing Data Use the periodic table to determine the molar mass of NaCl. Then calculate the number of moles of NaCl in the evaporating dish.

$$31.4 \text{ g} \times \frac{1 \text{ mol NaCl}}{58.44 \text{ g}} = 0.0537 \text{ mol NaCl}$$

3. Analyzing Data Calculate the molar concentration of the 10.0 mL of NaCl solution that you evaporated.

$$\frac{0.0537 \text{ mol NaCl}}{10.0 \text{ mL}} \times \frac{1000 \text{ mL}}{1 \text{ L}} = 5.37 \text{ M}$$

4. Organizing Ideas Write the chemical equation for dissolving NaCl to form the saturated solution.

The equilibrium equation for dissolving NaCl to form a saturated solution is
$NaCl(s) \rightleftharpoons Na^+(aq) + Cl^-(aq)$.

| Solubility Product Constant *continued*

5. Organizing Data Use the concentration from item **3** and the equation from item **4** to determine the concentrations of Na^+ and Cl^- ions in the saturated solution.

The equation in item 4 shows a 1:1 ratio between NaCl and Na^+ and between NaCl and Cl^-. NaCl is a strong electrolyte, which dissolves completely. Therefore, $[Na^+] = [Cl^-] = 5.37$ M.

6. Applying Ideas Write the equilibrium expression for the equation in item **4**.
$K_{sp} = [Na^+][Cl^-]$

7. Analyzing Data Use your answers to items **5** and **6** to calculate the numerical value of the K_{sp} for NaCl.
$K_{sp} = (5.37)(5.37) = 28.8$

8. Organizing Data How many moles of $Na_3C_6H_5O_7 \cdot 2H_2O$ were added to the solution in **step 11**?

0.49 g $Na_3C_6H_5O_7 \cdot 2H_2O \times \dfrac{1 \text{ mol}}{294.12 \text{ g}} = 1.7 \times 10^{-3}$ mol

Conclusions

9. Applying Conclusions Use your results from item **3** to determine the mass of NaCl needed to make 4000.0 L of saturated solution.

4000.0 L $\times \dfrac{5.37 \text{ mol NaCl}}{1 \text{ L}} \times \dfrac{58.44 \text{ g}}{1 \text{ mol}} = 1.26 \times 10^6$ g (or 1260 kg)

10. Analyzing Information Using the equilibrium expression from item **6**, explain your observations from **step 11**, when $Na_3C_6H_5O_7 \cdot 2H_2O$ was added to the NaCl solution.

The addition of $Na_3C_6H_5O_7$ to a saturated solution of NaCl increases the

concentration of Na^+. Any increase in concentration of either ion causes the

numerical value of K_{sp} to be exceeded. Therefore, NaCl precipitates from

the solution until the product of the concentrations of the ions again equals

the K_{sp}.

11. **Predicting Outcomes** The development team at Juliette Brand Foods is also considering adding the following flavoring or preservative additives to the brine: acetic acid, aluminum sulfate, calcium chloride, sodium bicarbonate, and sodium nitrate. Indicate which ones will cause a change in the NaCl equilibrium concentration.

The additives that will exhibit the common-ion effect with a NaCl solution

are the ones most likely to have an effect. These are calcium chloride,

sodium bicarbonate, and sodium nitrate.

Extensions

1. **Applying Conclusions** Suppose your client diluted the saturated NaCl brine to one half of its molarity. Would it be possible then to add 50.0 g of $CaCl_2$ to every liter of the diluted solution without causing precipitation? Show calculations to support your answer.

If the NaCl brine is diluted to half of its concentration, $[Na^+] = [Cl^-] = 2.68$ M.

$$\frac{50.0 \text{ g } CaCl_2}{1 \text{ L}} \times \frac{1 \text{ mol } CaCl_2}{110.98 \text{ g}} = 0.451 \text{ M}$$

$[Cl^-]$ from $CaCl_2 = 0.902$ M

total $[Cl^-] = 2.68$ M $+ 0.902$ M $= 3.58$ M

Using the equilibrium expression, $[Na^+][Cl^-] = (2.68)(3.58) = 9.59$. Because K_{sp} (28.8 at this temperature) is not exceeded, no precipitation will occur.

2. **Applying Ideas** Although iron(III) nitrate and iron(III) acetate have very large solubility product constants, farmers do not find these compounds useful as additives for iron-deficient soils. Explain why this is true.

The large solubility product constants of iron(III) nitrate and iron(III)

acetate mean that they are very soluble. These compounds are too quickly

removed by rain and irrigation to be useful for boosting the iron content of

soil.

| Solubility Product Constant *continued*

3. Applying Ideas Assuming that the value for K_{sp} of NaCl stays the same and that all of the $Na_3C_6H_5O_7 \cdot 2H_2O$ dissolves, you can calculate the new equilibrium concentration for Na^+ and Cl^- using the following steps.
- Calculate the total possible concentration of Na^+ ions.
- Assume that x mol/L of NaCl precipitates.
- Rewrite the equilibrium expression with $[Na^+]$ equal to the total possible concentration minus x and $[Cl^-]$ equal to the initial concentration minus x.
- Algebraically rearrange the equation until there is a zero on one side of the equation.
- Apply the quadratic formula to the equation from the previous step to determine the value of x, the equilibrium concentration of NaCl. For any equation with the form given below, x has the values shown.

$$ax^2 + bx + c = 0$$

$$x = \frac{-b \pm \sqrt{b^2 - 4ac}}{2a}$$

$[Na^+]$ from $Na_3C_6H_5O_7 = 3 \times \dfrac{1.7 \times 10^{-3} \text{ mol}}{0.005 \text{ L}} = 1.02 \text{ M}$

total possible $[Na^+]$ = 5.37 M + 1.02 M = 6.39 M

Assuming that x mol/L of Na^+ and Cl^- ions will precipitate as solid NaCl, the equilibrium expression has the following values.

$K_{sp} = 28.8 = [Na^+][Cl^-] = (6.39 - x)(5.37 - x)$

$0 = x^2 - 11.76x + 5.5$

$x = \dfrac{\sqrt{11.76 \pm (-11.76)^2 - 4(5.5)}}{2}$

x = 0.5 or 11.3 (11.3 is impossible because that much was not originally available)

$[Na^+]$ = 5.89 M

$[Cl^-]$ = 4.87 M

Solubility Product Constant *continued*

4. **Research and Communication** The hollowing out of limestone caves and the growth of stalactites and stalagmites are environmental examples of solubility equilibria at work. Investigate these processes in terms of equilibrium, and prepare a paper explaining the formation of caves.

Student answers will vary, but students should see that the equilibrium

equation that follows is a key part of the process.

$$CaCO_3(s) + H_2O(aq) + CO_2(g) \rightleftharpoons Ca^{2+}(aq) + 2HCO_3^-(aq)$$

When calcium carbonate is exposed to air, a small amount of it reacts with

and dissolves in water due to CO_2 in the air, forming the aqueous solution

indicated on the right side of the equation. After the water seeps into a cave,

which typically has a lower amount of CO_2 in its air, the reverse reaction

occurs, leaving a layer of insoluble $CaCO_3$ behind.

5. **Applying Ideas** The method for producing sweet pickles is essentially the same as that for making dill pickles, except that the cucumbers are soaked in a saturated sucrose solution instead of a NaCl solution. (Incidentally, sweet pickles were invented when the person doing the pickling mistakenly put sugar into the water instead of salt.) Depending upon the success of their new line of dill pickles, Juliette Brand Foods is planning to expand production to include sweet pickles. Based on your knowledge of how sugar dissolves in water, would you be able to experimentally determine a solubility product constant for sucrose? Does this affect the precipitation of sugar in the presence of a common ion? Explain your answers.

Although the solubility limits of sucrose can be determined experimentally,

sucrose does not dissociate. Thus, a solubility product constant cannot be

determined and there should be no common-ion effect.

| **Solubility Product Constant** *continued*

6. **Predicting Outcomes** Because they are just beginning to enter this market, Juliette Brand Foods does not want to have to spend more capital than necessary on equipment. Included in this would be a temperature control for the brine vats, so that the solution remains at a constant temperature at all times. Does the solubility of NaCl vary sufficiently over a range of temperatures to make such a temperature control device necessary? The dependence of solubility on temperature for several solutes is shown in the textbook. Use this information to infer how the solubility product constant for NaCl varies with temperature.

The solubility of NaCl changes very little with temperature, so a tempera-

ture control device should be unnecessary. K_{sp} should not change much.

7. **Relating Ideas** Often it is convenient to express the solubility product constant as pK_{sp}, which is defined similarly to pH.

$$pK_{sp} = -\log K_{sp}$$

From this definition, what is the value of K_{sp} for a pK_{sp} value of 2.5? 25.7?

$pK_{sp} = -\log K_{sp}$

$2.5 = -\log (3 \times 10^{-3})$ $25.7 = -\log (2 \times 10^{-26})$

Solubility Product Constant–Algal Blooms

Teacher Notes

TIME REQUIRED One to two 45-minute lab periods

SKILLS ACQUIRED
Collecting data
Identifying/Recognizing patterns
Interpreting
Measuring
Organizing and analyzing data

RATING
Easy ←— 1   2   3   4 —→ Hard

Teacher Prep–2
Student Set-Up–3
Concept Level–2
Clean Up–1

THE SCIENTIFIC METHOD

Make Observations

Analyze the Results

Draw Conclusions

MATERIALS
(for each lab group)
- balance
- beakers, 150 mL, (4)
- crucible tongs
- $CuCl_2 \cdot 2H_2O$ (25 g)
- $CuSO_4 \cdot 5H_2O$ (15 g)
- distilled water
- graduated cylinder, 10 mL
- evaporating dishes (2)
- glass stirring rods (2)
- spatula

Bunsen burner option
- Bunsen burner
- gas tubing
- ring stand and ring
- striker
- wire gauze with ceramic center

Hot plate option
- hot plate

Estimated cost of materials:
$117 000–$132 000

SAFETY CAUTIONS

Safety goggles and a lab apron must be worn at all times.

Tie back long hair and loose clothing.

Read all safety cautions, and discuss them with your students.

Remind students that heated objects can be hot enough to burn even if they look cool. Students should always use crucible tongs when handling any lab equipment that has been heated. If a hotplate is used, precautions must be followed to avoid electric shock.

Remind students of the following safety precautions:

- Always wear safety goggles, gloves, and a lab apron to protect your eyes and clothing. If you get a chemical in your eyes, immediately flush the chemical out at the eyewash station while calling to your teacher. Know the location of the emergency lab shower and the eyewash stations and procedures for using them.

- Do not touch any chemicals. If you get a chemical on your skin or clothing, wash the chemical off at the sink while calling to your teacher. Make sure you carefully read the labels and follow the precautions on all containers of chemicals that you use. If there are no precautions stated on the label, ask your teacher what precautions you should follow. Do not taste any chemicals or items used in the laboratory. Never return leftovers to their original containers; take only small amounts to avoid wasting supplies.

- Call your teacher in the event of a spill. Spills should be cleaned up promptly, according to your teacher's directions.

- Never put broken glass in a regular waste container. Broken glass should be disposed of properly.

DISPOSAL

Provide two labeled containers for the disposal of $CuSO_4$ and $CuCl_2$ solutions and any excess of the original compounds. Later, redissolve the contents of the container in distilled water. Let the solution evaporate until dry, and then recover the crystals for re-use next year. Do not dispose of these compounds in a landfill or incinerator or down the drain.

TECHNIQUES TO DEMONSTRATE

Show students how to gently heat the saturated solutions to avoid spattering. When the solutions appear dry, students should continue to heat them gently until $CuSO_4$ is white and $CuCl_2$ is light brown. If a yellow color persists in the $CuSO_4$ after it has cooled, that is a sign that the heating was too rapid and decomposition occurred.

TIPS AND TRICKS

Begin by discussing the results and procedure used in the Solubility Product Constant, especially any errors in lab technique that occurred. Make sure they understand the answers to the Conclusions for the Solubility Product Constant.

Some students may decide they don't need to do this lab at all because the K_{sp} values are available in reference books. Tell them that students need data to support their conclusions because the K_{sp} values from reference sources are obtained under ideal conditions with very pure reagents.

Be certain that students have developed a procedure that details exactly how they will divide the lab work. For instance, one member of the group can be preparing one solution and evaporating it while another member prepares the other solution. Otherwise, this lab will be difficult to finish in time.

ANSWERS

Procedure

Place 25 mL of distilled water in a 150 mL beaker. Add 15 g of $CuCl_2 \cdot 2H_2O$, and stir until it dissolves. Continue to add salt in 5 g increments, while stirring, until some remains undissolved on the bottom. Repeat the procedure for $CuSO_4 \cdot 5H_2O$ in another 150 mL beaker.

Decant the saturated solutions into two other labeled beakers.

Measure and record the masses of two labeled evaporating dishes.

Measure 10.0 mL of each solution into separate evaporating dishes, and record the volumes of the solutions, washing the graduated cylinder between uses.

Using a Bunsen burner or hot plate, gently evaporate the water from each solution in the evaporating dishes. Continue heating until the salts are dry and anhydrous.

Allow the evaporating dishes to cool and obtain the masses of the evaporating dishes and salts.

Estimated cost of materials: $86000 ($101 000 with spectroscope)

Sample Results

	$CuSO_4$	$CuCl_2$
Mass of evaporating dish	41.36 g	41.36 g
mL of saturated solution	10.0 mL	10.0 mL
Mass of evaporating dish and salt	43.21 g	47.06 g

For both salts

$$3.90 \times 10^6 \text{ L} \times \frac{0.0500 \text{ mol}}{1 \text{ L}} = 1.95 \times 10^5 \text{ mol of salt needed for a pond}$$

For $CuSO_4$

Molarity of solution:

$$\frac{1.85 \text{ g}}{10.0 \text{ mL}} \times \frac{1 \text{ mol CuSO}_4}{159.62 \text{ g}} \times \frac{1000 \text{ mL}}{1 \text{ L}} = 1.16 \text{ M}$$

$K_{sp} = [Cu^{2+}][SO_4^{2-}] = (1.16)^2 = 1.35$

Volume of saturated solution needed:

$$(1.95 \times 10^5 \text{ mol}) \times \frac{1 \text{ L}}{1.16 \text{ mol}} \times 1.68 \times 10^5 \text{ L}$$

Mass of hydrated salt needed:

$$(1.95 \times 10^5 \text{ mol}) \times \frac{249.72 \text{ g}}{1 \text{ mol}} \times \frac{1 \text{ kg}}{1000 \text{ g}} = 4.87 \times 10^4 \text{ kg}$$

Cost: $(4.87 \times 10^7 \text{ g}) \times \dfrac{\$100}{1 \text{ g}} = \$4.87 \times 10^9$

For CuCl$_2$

Molarity of solution:

$$\frac{5.70 \text{ g}}{10.0 \text{ mL}} \times \frac{1 \text{ mol CuCl}_2}{134.45 \text{ g}} \times \frac{1000 \text{ mL}}{1 \text{ L}} = 4.24 \text{ M}$$

$K_{sp} = [\text{Cu}^{2+}][\text{Cl}^-]^2 = (4.24)(8.48)^2 = 305$

Volume of saturated solution needed:

$$(1.95 \times 10^5 \text{ mol}) \times \frac{1 \text{ L}}{4.24 \text{ mol}} = 4.60 \times 10^4 \text{ L}$$

Mass of hydrated salt needed:

$$(1.95 \times 10^5 \text{ mol}) \times \frac{170.49 \text{ g}}{1 \text{ mol}} \times \frac{1 \text{ kg}}{1000 \text{ g}} = 3.32 \times 10^4 \text{ kg}$$

Cost: $(3.32 \times 10^7 \text{ g}) \times \dfrac{\$200}{1 \text{ g}} = \$6.64 \times 10^9$

Students may chose CuCl$_2 \cdot$2H$_2$O as the best choice to kill algae because less mass and smaller volumes of saturated solution are needed. Alternatively, CuSO$_4 \cdot$5H$_2$O costs less per gram. A mixture will not provide a greater concentration because of the common-ion effect.

Inquiry

Solubility Product Constant–Algal Blooms

April 18, 2003

Ms. Sandra Fernandez

Director of Development

CheMystery Labs, Inc.

52 Fulton Street

Springfield, VA 22150

Dear Ms. Fernandez:

Recently, our region has experienced above-average temperatures and rainfall. These factors have increased runoff into local waterways, causing abnormal algae blooms in lakes and ponds. Our studies show that algae can be controlled with 0.0500 M solutions of copper(II) ions, so we are considering treating affected ponds with copper(II) sulfate or copper(II) chloride.

We would like to apply them to the ponds and lakes in the form of concentrated solutions. In this way, they will mix thoroughly with the lake or pond water much more quickly than if we added the solid compounds.

We are requesting bids for several comparative studies of the solubility properties of these two compounds. We want each contractor to recommend one compound or an optimum combination of both for our use. Please base any calculations of amounts necessary on a 3.90×10^6 L pond and submit experimental data to support your conclusions.

Sincerely,

Kathleen Farros-Hoeppner

Assistant Director, Research

State Department of Fish and Game

References

Reread the sections of your textbook that discuss information about saturated solutions and molarity. Review the background information about equilibrium systems, and the equilibrium constant.

Hint: because these compounds can form hydrates, $CuCl_2 \cdot 2H_2O$ and $CuSO_4 \cdot 5H_2O$, they should be heated thoroughly to produce the anhydrous compounds after the solvent is evaporated. $CuSO_4$ should be gray or white (not blue or yellow), and $CuCl_2$ should be light brown.

| Solubility Product Constant—Algal Blooms *continued*

CheMystery Labs, Inc.
52 Fulton Street
Springfield, VA 22150

Memorandum

Date: April 19, 2003

To: Gary Vasileyev

From: Sandra Fernandez

I think the best way to proceed is to measure the concentrations of saturated solutions of each compound. Make a 25 mL saturated solution of each one and calculate K_{sp} so that our results can be compared quickly to those of other firms.

If we land one of these contracts, it will be our first project for a state agency, so we want to make a favorable first impression. Send Ms. Farros-Hoeppner a bid that includes the following.

- detailed one-page plan for the procedure
- examples of all necessary data tables
- list of the materials you need with the total cost

Get started as soon as your plan is approved. When the work is complete, prepare a report in the form of a two-page letter to Kathleen Farros-Hoeppner. The letter must include the following items.

- your recommendation of which compound to kill the algae (be sure to consider costs of each one)
- volumes of each saturated solution needed to achieve a concentration of 0.05 M Cu(II) ions in the pond
- mass of each compound needed to make the solutions
- equilibrium molar concentrations of saturated solutions
- experimentally measured K_{sp} of copper(II) sulfate and copper(II) chloride based on experimental data
- brief description of the procedure
- detailed and organized data table and calculations
- detailed invoice for materials and services

Solubility Product Constant—Algal Blooms *continued*

Always wear safety goggles and a lab apron to protect your eyes and clothing. If you get a chemical in your eyes, immediately flush the chemical out at the eyewash station while calling to your teacher. Know the location of the emergency lab shower and eyewash station and the procedures for using them.

Do not touch any chemicals. If you get a chemical on your skin or clothing, wash the chemical off at the sink while calling to your teacher. Make sure you carefully read the labels and follow the precautions on all containers of chemicals that you use. If there are no precautions stated on the label, ask your teacher what precautions to follow. Do not taste any chemicals or items used in the laboratory. Never return leftovers to their original container; take only small amounts to avoid wasting supplies.

Do not heat glassware that is broken, chipped, or cracked. Use tongs or a hot mitt to handle heated glassware and other equipment because hot glassware does not always look hot.

When using a Bunsen burner, confine long hair and loose clothing. If your clothing catches on fire, WALK to the emergency lab shower and use it to put out the fire. Do not heat glassware that is broken, chipped, or cracked. Use tongs or a hot mitt to handle heated glassware and other equipment because hot glassware does not always look hot.

In case of spills, follow your teacher's instruction. Dispose of solid wastes in the waste container designated by your teacher. Dispose of liquid wastes in the waste container designated by your teacher. Always wash your hands thoroughly when finished.

❚ Solubility Product Constant—Algal Blooms *continued*

MATERIALS FOR STATE DEPARTMENT OF FISH AND GAME

Required Items (You must include all of these in your budget.)	
Lab space/fume hood /utilities	15,000 /day
Standard disposal fee	2,000 /g of product
Balance	5,000
Crucible tongs	2,000
Evaporating dish	1,000

Reagents and Additional Equipment (Include in your budget only what you'll need.)	
$CuCl_2 \cdot 2H_2O$	200 /g
$CuSO_4 \cdot 5H_2O$	100 /g
150 mL beaker	1,000
400 mL beaker	2,000
250 mL flask	1,000
25 mL graduated cylinder	1,000
100 mL graduated cylinder	1,000
Filter paper	500 /piece
Glass stirring rod	1,000
Bunsen burner/related equipment	10,000
Hot plate	8,000
Litmus paper	1,000 /piece
Ring stand/ring/wire gauze	2,000
Ruler	500
Spatula	500
Test tube (large)	1,000
Test tube (small)	500
Wash bottle	500
* No refunds on returned chemicals or unused equipment.	

Fines	
OSHA safety violation	2,000 /incident

Rust Race

Teacher Notes

TIME REQUIRED Day 1: 60 min to set up investigations, Day 4: 45 min to observe and record results

SKILLS ACQUIRED
Collecting data
Communicating
Designing Experiments
Experimenting
Identifying patterns
Inferring
Interpreting
Organizing and analyzing data

RATING
Easy ←——1——2——3——4——→ Hard

Teacher Prep–3
Student Set-Up–3
Concept Level–3
Clean Up–3

THE SCIENTIFIC METHOD

Make Observations Students observe rust formation under various conditions.

Analyze the Results Analysis questions 1 to 4 and 6 to 7

Draw Conclusions Analysis question 5 and Conclusions questions 1 and 2

Communicate the Results Analysis questions 1 to 7

MATERIALS

To prepare 1 L of a 2% salt solution, add 20 g of sodium chloride (table salt), NaCl, to 1 L of tap water.

To prepare 1 L of a vinegar solution, mix 250 mL of store-bought vinegar with 750 mL of tap water.

To prepare bleach solution for washing the steel wool, mix one part bleach to nine parts water. Be sure to observe proper safety precautions when using bleach.

Obtain any commercially available rust inhibitor from a hardware store.

The nails, paper clips, quarters, and pennies should initially be shiny.

SAFETY CAUTIONS

Read all safety precautions, and discuss them with your students. Include precautions on labels of commercially available products.

- Safety goggles and a lab apron must be worn at all times.

- In case of a spill, use a dampened cloth or paper towels to mop up the spill. Then rinse the towels in running water at the sink, wring them out until only damp, and put them in the trash.

- Broken glass should be disposed of in a clearly labeled box lined with a plastic trash bag. When the box is full, close it, seal it with packaging tape, and set it next to the trash can for disposal.

- Students should avoid direct contact with the bleach solution. If contact with this solution occurs, the affected areas should be thoroughly rinsed with water.

TIPS AND TRICKS

To save laboratory time, wash the steel wool before the lab begins. Assist students with coating the nails.

Students should understand the concepts of oxidation and reduction. Review the activity series of metals and its relationship to oxidation. Discuss the results of **Data Table 3** in a post-lab discussion.

DISPOSAL

Set up a disposal container for any sharp objects such as nails and other metal objects. Used vinegar and salt solutions may be poured down the drain. Dispose of all other waste materials according to the directions provided by the manufacturer of the product.

Inquiry

Rust Race

When a metal corrodes, it oxidizes, or transfers electrons to a nonmetal, such as oxygen. Copper reacts with oxygen in the air to form the soft green substance called verdigris. When iron is exposed to moist air, it reacts with oxygen to form the reddish brown substance iron oxide, or rust.

Other factors, such as temperature, the presence of salt, acidity, and air or water pollutants, also affect the rate of metal corrosion. For example, iron and steel tend to corrode much more quickly when exposed to salt, such as that used to melt snow, or to a moist, salty environment, such as in areas near the ocean. This is because dissolved salts increase the conductivity of the aqueous solution formed at the surface of the metal.

A metal can be protected from corrosion by being coated with paint, rust inhibitor, or a special film. These coatings prevent exposure of the metal to oxygen or moisture. Galvanizing is the most widely used method to protect iron and steel products from corrosion. During this process, a thin coating of zinc is applied to a metal. When the galvanized metal is exposed to elements in the atmosphere, the zinc coating oxidizes instead of the iron or steel.

In this experiment, you will test metals for oxidation, measure oxygen consumption during metal corrosion, and identify ways to protect iron metal from corrosion.

OBJECTIVES

Test metals for corrosion.

Measure oxygen consumption during metal corrosion.

Compare the effectiveness of rust inhibitors.

MATERIALS

- bleach solution
- clear nail polish
- galvanized nails, 3
- iron nails, 8
- paper clips, 3
- pennies, 3
- quarters, 3
- petroleum jelly
- rust inhibitor, commercially available
- 2% salt solution
- steel wool, plain
- stirring rod
- vinegar solution
- 100 mL beakers, 6
- 600 mL beaker
- 50 mL graduated cylinder
- balance
- forceps
- metric ruler
- shredded paper
- test tubes (16 × 125 mm), 2
- water
- wax pencil
- plastic wrap

Always wear safety goggles and a lab apron to protect your eyes and clothing. If you get a chemical in your eyes, immediately flush the chemical out at the eyewash station while calling to your teacher. Know the locations of the emergency lab shower and the eyewash station and the procedures for using them.

Do not touch any chemicals. If you get a chemical on your skin or clothing, wash the chemical off at the sink while calling to your teacher. Make sure you carefully read the labels and follow the precautions on all containers of chemicals that you use. If there are no precautions stated on the label, ask your teacher what precautions you should follow. Do not taste any chemicals or items used in the laboratory. Never return leftovers to their original containers; take only small amounts to avoid wasting supplies.

Call your teacher in the event of a spill. Spills should be cleaned up promptly, according to your teacher's directions.

Never put broken glass in a regular waste container. Broken glass should be disposed of properly.

Procedure
PART 1: TESTING METALS FOR CORROSION

1. Use a wax pencil to label three 100 mL beakers *water*, *salt*, and *vinegar*. Using a graduated cylinder, pour into each beaker 50 mL of the solution corresponding to the beaker's label. Rinse the graduated cylinder with deionized water after adding salt solution and vinegar solution.

2. Using forceps, place the following uncorroded objects in each beaker: a penny, a quarter, an iron nail, a paper clip, and a galvanized nail. Cover the beakers with plastic wrap. Then place the three beakers in an undisturbed area of the laboratory, and let them stand undisturbed for four or five days.

3. After four or five days, use forceps to remove your test materials from each beaker, laying the contents of each beaker on a separate paper towel. Carefully observe each object for any physical changes that may have taken place. Record any signs of metal corrosion (loss of shine, change in color, rust or verdigris formation) in **Data Table 1.**

| Rust Race *continued*

Data Table 1

Test material	In water			In salt solution			In vinegar solution		
	Shine	Color	Rust	Shine	Color	Rust	Shine	Color	Rust
Iron nail	none	reddish brown	yes	none	reddish brown	yes	none	reddish brown	yes
Galvanized nail	duller	silverish gray	no	duller	silverish gray	no	duller	silverish gray	no
Paper clip	duller	silverish gray	no	duller	silverish gray	no	duller	silverish gray	no
Quarter	duller	silverish gray	no	duller	silverish gray	no	duller	silverish gray	no
Penny	less luster	copper/ greenish	no	none	greenish	no	none	greenish	no

PART 2: MEASURING OXYGEN CONSUMPTION DURING METAL CORROSION

4. Use a wax pencil to label two test tubes *steel wool* and *control*.

5. Measure the mass of about 1 g of steel wool to the nearest 0.1 g. Record the mass in **Data Table 2.** To remove any rust inhibitors it may contain, place the steel wool into a 100 mL beaker. Pour in bleach to cover, and stir with a stirring rod. This ensures that oxygen can access the surface of the metal for oxidation to take place.

6. Rinse the bleach out of the steel wool, and pull apart the strands so that they are loosely packed and fluffy. Fill the bottom third of the *steel wool* test tube with the loosely packed steel wool. The steel wool should make contact with the sides of the test tube and should not slide out when the tube is inverted.

7. Measure the mass of about 1 g of shredded paper to the nearest 0.1 g. Record the mass in **Data Table 2.** Fill the bottom third of the *control* test tube with the shredded paper. It should make contact with the sides of the test tube and should not slide out when the tube is inverted.

8. Pour about 200 mL of water into a 600 mL beaker. Invert the test tubes from steps **6** and **7,** and place them open-end down in the beaker of water. The test tubes should rest against the side of the beaker. Keep the tubes' open ends fully beneath the water level.

9. Use a wax pencil to mark the water level in each test tube. (It is possible that no water will enter the tube initially. In that case, simply record height of water at the start as 0.) Use a metric ruler to measure the height of water in each tube. Also measure the height of the test tube above the water. Record your measurements in **Data Table 2.** Leave your setup undisturbed for 24 hours.

Rust Race *continued*

10. After 24 hours, mark the new water level in each test tube. Use a metric ruler to measure the new water level in each test tube. Record your results in **Data Table 2.** Why did the water rise in the test tube with the steel wool but not in the test tube with water?

    The water rises to fill the space left behind when oxygen molecules in the air

    above the water combine with the steel wool to form rust. No reaction takes

    place between the paper and oxygen, so the water does not rise into the

    paper.

11. Allow the test tubes containing the steel wool and paper to sit vertically, open-end up, for at least 2 days. This allows the steel wool to dry.

12. Using forceps, remove the steel wool from the test tube and measure its mass. Record your results in **Data Table 2.** Using forceps, remove the shredded paper from the test tube and measure its mass. Record your results in **Data Table 2.** Carefully observe the steel wool strands and the shredded paper. Describe any corrosion that has taken place.

    Observations:

    Answers will vary. _____

    _____

    _____

    _____

Data Table 2

Measurement	Steel wool	Shredded paper
Height of test tube above water level (cm)	7.5	7.5
Height of water at start (cm)	5	5
Height of water after 24 h (cm)	6.5	same as initial height
Water level increase (cm)	1.5	0
Percent increase in water level (%)	20	0
Percent of oxygen consumed, by volume (%)	20	0
Initial mass (g)	1.0	1.0
Final mass (g)	1.1	1.0
Change in mass (g)	0.1	0.0

Note: Height of the water and test tube will vary depending on the setup used.

Rust Race *continued*

PART 3: PROTECTING METALS FROM CORROSION

13. Obtain five iron nails. Thoroughly coat the entire surface of the first nail with petroleum jelly, coat the second nail with a commercially available rust inhibitor, coat the third nail with clear nail polish, and coat the fourth nail with a coating of your choice. Leave the fifth nail uncoated; it is your control. Allow the nail polish and rust inhibitor to dry. To ensure that you know which nail has which coating, make a map or sketch of their locations in your lab notes.

14. Pour 50 mL of salt solution in a 100 mL beaker. Using forceps, place your coated nails in the beaker. Be sure the nails do not touch each other. Make another map or sketch of the nails' locations. Place the beaker containing the nails in an undisturbed area of the laboratory for four or five days.

15. After four or five days, use forceps to remove the nails from the beaker. Lay the nails on separate paper towels. Carefully observe each nail for any physical changes that may have taken place. Record any signs of metal corrosion (loss of shine, change in color, or rust formation) in **Data Table 3.**

Data Table 3

Test material	Observations		
	Shine	**Color**	**Rust**
Iron nail (control)	less luster	reddish brown	yes
Iron nail coated with petroleum jelly	N/A	original color	no
Iron nail coated with rust inhibitor	original luster	original color	no
Iron nail coated with clear nail polish	none	original color	yes, some
Iron nail with coating of your choice	variable, depending on the coating		

DISPOSAL

16. Clean all apparatus and your lab station. Return equipment to its proper place. Dispose of chemicals and solutions in the containers designated by your teacher. Do not pour any chemicals down the drain or put them in the trash unless your teacher directs you to do so. Wash your hands thoroughly after all work is finished and before you leave the lab.

Rust Race *continued*

Analysis

Refer to the results of Part 2 of the experiment to answer questions 1–5.

1. **Organizing Data** Calculate the water level increase for each test tube by subtracting the initial height of the water from the height of the water after 24 hours. Record your results in **Data Table 2.**

2. **Organizing Data** Calculate for both test tubes the percentage increase in water level by dividing the increase in water level by the height of the test tube above the water, then multiply the answer by 100. Record your results in **Data Table 2.**

3. **Organizing Data** Calculate the change in the mass of the steel wool by subtracting the initial mass of the steel wool from its final mass. Record your result in **Data Table 2.**

   Students should note a mass increase of 0.1 g or more, depending on the extent of corrosion that has taken place.

4. **Organizing Data** Calculate the change in the mass of the shredded paper by subtracting the initial mass of the paper from its final mass. Record your result in **Data Table 2.**

   Students should note a mass increase of 0.0 g.

5. **Inferring Conclusions** Based on your calculations in item **2**, what is the percentage of oxygen consumed, by volume, in each test tube?

   The water in the tube will rise by the same percentage as the oxygen con-

   sumed. Because oxygen makes up about 20% of the composition of air,

   students should observe a maximum of 20% increase in water level.

Rust Race *continued*

6. **Analyzing Data** According to your observations in **Data Table 1,** which solution(s) caused the most corrosion in the materials tested?

Salt and/or vinegar solutions caused the most corrosion; students' data

should support their responses.

7. **Analyzing Data** Explain why the steel wool gained mass in **Part 2** of the investigation.

The steel wool combined with oxygen from the air and produced iron oxide

(rust).

Conclusions

1. **Designing Experiments** The tarnishing of silver is an oxidation reaction between silver and sulfur in the air or in certain foods. Identify a food that contains sulfur, and design an experiment to demonstrate the tarnishing of silver.

Answers will vary. One possibility is to push part of a shiny silver coin into

the egg white of a hard-boiled egg. Remove the coin after an hour. Because

the egg contains sulfur, it will react with silver to form a tarnish on the

portion of the coin covered by the egg white.

2. **Designing Experiments** Design an experiment to determine which brand of steel wool resists rusting the longest.

Answers will vary. One possible solution is to place 1 g samples of each steel

wool to be tested into separate test tubes. Add 10 mL of vinegar to each test

tube. Let the setups sit undisturbed. Observe the samples periodically, and

record the extent of corrosion in each case.

Electroplating for Corrosion Protection

Teacher Notes

TIME REQUIRED One 45-minute lab period with drying oven; two 45-minute lab periods otherwise

SKILLS ACQUIRED

Collecting data
Communicating
Identifying patterns
Inferring
Interpreting
Organizing and analyzing data

RATING

Easy ←——|——|——|——|——→ Hard
               1   2   3   4

Teacher Prep–3
Student Set-Up–3
Concept Level–2
Clean Up–2

THE SCIENTIFIC METHOD

Make Observations Students will observe and compare the results of electroplating copper wire with iron and with zinc.

Analyze the Results Analysis question 2 and Conclusions questions 3 and 4

Draw Conclusions Conclusions questions 3 and 4.

Communicate the Results Analysis question 2 and Conclusions questions 3 and 4

MATERIALS

To prepare 1.00 L of the $FeCl_3$ plating solution, observe the required precautions. Dissolve 40.0 g $FeCl_3 \cdot 6H_2O$ in 900 mL of distilled water and 50 mL of ethanol. Add more water to dilute to 1.00 L.

To prepare 1.00 L of the $ZnSO_4$ plating solution, dissolve 50.0 g $ZnSO_4 \cdot 7H_2O$, 24.0 g NH_4Cl, and 40.0 g ammonium citrate, $(NH_4)_2HC_6H_5O_7$, in 900 mL of distilled water. Add more water to dilute to 1.00 L.

To prepare 1.00 L of 1.00 M HCl, observe the required precautions. Add 82.6 mL of concentrated HCl to enough distilled water to make 1.00 L of solution. Add the acid slowly, and stop frequently to stir it in order to avoid overheating.

Any size of copper wire will do. The procedure that generated the sample data was performed with soft, bare, 18 gauge wire.

Use a 6 V "lantern battery" or its equivalent. DO NOT use a high-amperage 6 V battery such as those designed for supplying current for the spark in gasoline-powered motors.

Although optional, the use of the drying oven is strongly recommended. Otherwise, if the lab is to be performed quantitatively, the wires must dry overnight after plating and again after testing them in 1.0 M HCl.

SAFETY CAUTIONS

Remind students that objects from the drying oven will be very hot. They should use beaker tongs or a hot mitt to pick them up.

Students should not handle concentrated acid or base solutions.

Wear safety goggles, a face shield, impermeable gloves, and a lab apron when preparing the HCl. Work in a hood known to be in operating condition, with another person present nearby to call for help in case of an emergency. Be sure you are within 30 s walk of a properly working safety shower and eyewash station.

In case of an acid or base spill, first dilute with water. Then mop up the spill with wet cloths or a wet cloth mop while wearing disposable plastic gloves. Designate separate cloths or mops for acid and base spills.

Observe the precautions on the ethanol bottle's label when preparing the $FeCl_3$ solution.

Remind students of the following safety precautions:

- Always wear safety goggles, gloves, and a lab apron to protect your eyes and clothing. If you get a chemical in your eyes, immediately flush the chemical out at the eyewash station while calling to your teacher. Know the location of the emergency lab shower and the eyewash station and the procedure for using them.

- Do not touch any chemicals. If you get a chemical on your skin or clothing, wash the chemical off at the sink while calling to your teacher. Make sure you carefully read the labels and follow the precautions on all containers of chemicals that you use. If there are no precautions stated on the label, ask your teacher what precautions you should follow. Do not taste any chemicals or items used in the laboratory. Never return leftovers to their original containers; take only small amounts to avoid wasting supplies.

- Call your teacher in the event of a spill. Spills should be cleaned up promptly, according to your teacher's directions.

- Never put broken glass in a regular waste container. Broken glass should be disposed of properly.

DISPOSAL

Set out six disposal containers: three bins and three bottles, or other similar containers. Of the three bins, use one for copper wires, one for zinc metal strips, and one for iron metal strips. Of the three containers, use one for the $FeCl_3$ solution, one for the $ZnSO_4$ solution, and one for the contents of the *Waste* beaker and the HCl from the test tubes.

The metal strips should be cleaned with soap and water, rinsed, and dried for reuse the next time students perform this lab. The copper wires can also be reused after they have been treated with 1.0 M HCl and all of the zinc or iron has been removed. The iron and zinc solutions should be saved and reused.

Pour the HCl that has been used in treating the copper wires into the disposal container for the *Waste* beaker. Add 1.0 M NaOH to the mixture in the *Waste* beaker container to precipitate the iron and zinc as hydroxides. Filter the mixture, placing the precipitate in the trash. Then neutralize the filtrate with 1.0 M H_3PO_4 until the pH is between 5 and 9, and pour it down the drain.

TECHNIQUES TO DEMONSTRATE

Be sure to show students the proper order for the connections between the electrodes and the battery, reminding them to check that they have connected the appropriate metal to the battery's positive electrode. Point out the importance of performing the steps in exactly the same manner for the different wire treatments. Each wire should be plated for the same amount of time, at the same depth in the beaker, and with similar amounts of plating solution. Similarly, the testing steps should be performed with the same amounts of acid for the same amount of time.

TIPS AND TRICKS

Discuss reduction and oxidation. Make sure students realize that a different reaction is occurring at each electrode. Many will recognize that plating (reduction) is occurring at one electrode but will not understand that oxidation of the metal is occurring at the other.

| Inquiry |

Electroplating for Corrosion Protection

SITUATION

Your company has been contacted by a manufacturer of electrical circuits. The company uses 1.0 M HCl to clean newly manufactured circuits. They've decided to store the acid in large metal tanks instead of reagent bottles. The company has narrowed down the choices to copper, zinc-plated copper, and iron-plated copper. You have been asked to evaluate these choices.

BACKGROUND

Solutions of acids can oxidize some metals. The displacement reaction of magnesium and hydrochloric acid is an example.

$$\underset{0}{Mg(s)} + \underset{+1\,-1}{2HCl(aq)} \longrightarrow \underset{+2\,-1}{MgCl_2(aq)} + \underset{0}{H_2(g)}$$

Metals that react can be electroplated with a thin layer of a less reactive metal. In an electroplating cell, electrical energy is used to reduce metal ions in solution, causing it to adhere to the surface of an object functioning as an anode. **Figure 1** shown here summarizes the apparatus and the process.

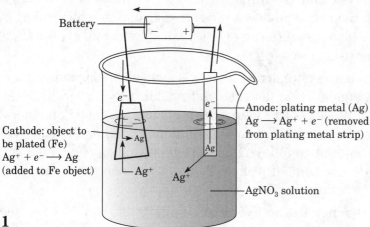

Figure 1

PROBLEM

You must first prepare samples of the plated metals. Then you must test each one in 1.0 M HCl to determine which resists corrosion the best.

OBJECTIVES

Construct an electrolytic cell with an electrolyte, two electrodes, and a battery.

Use the electrolytic cell to plate one metal onto another.

Test the plated metals and a sample of the original metal to determine how well they resist corrosion by an acidic solution.

Relate the results to the activity series and to standard reduction potentials.

Name _____ Class _____ Date _____

Electroplating for Corrosion Protection *continued*

MATERIALS

- balance
- battery, 6 V (lantern type)
- beaker, 400 mL
- beakers, 150 mL (3)
- copper wire, 10 cm lengths (3)
- distilled water
- FeCl$_3$ plating solution
- gloves
- HCl, 1.0 M
- iron strip, 1 cm × 8 cm
- lab apron
- safety goggles
- steel wool

- stick-on label
- stopwatch or clock with second hand
- test-tube rack
- test tubes, large (3)
- wax pencil
- wire with alligator clips (2 pieces)
- zinc strip, 1 cm × 8 cm
- ZnSO$_4$ plating solution

Optional equipment

- beaker tongs
- drying oven

  Always wear safety goggles, gloves, and a lab apron to protect your eyes and clothing. If you get a chemical in your eyes, immediately flush the chemical out at the eyewash station while calling to your teacher. Know the location of the emergency lab shower and eyewash station and the procedures for using them.

 Do not touch any chemicals. If you get a chemical on your skin or clothing, wash the chemical off at the sink while calling to your teacher. Make sure you carefully read the labels and follow the precautions on all containers of chemicals that you use. If there are no precautions stated on the label, ask your teacher what precautions to follow. Do not taste any chemicals or items used in the laboratory. Never return leftovers to their original container; take only small amounts to avoid wasting supplies.

Call your teacher in the event of a spill. Spills should be cleaned up promptly, according to your teacher's directions.

Acids and bases are corrosive. If an acid or base spills onto your skin or clothing, wash the area immediately with running water. Call your teacher in the event of an acid spill. Acid or base spills should be cleaned up promptly.

Do not heat glassware that is broken, chipped, or cracked. Use tongs or a hot mitt to handle heated glassware and other equipment because hot glassware does not always look hot.

Name _____ Class _____ Date _____

Electroplating for Corrosion Protection *continued*

PART 1–PREPARATION

1. Put on safety goggles, gloves, and a lab apron.

2. Label two of the 150 mL beakers *FeCl₃* and *ZnSO₄*. Label the 400 mL beaker *Waste*.

3. Make loops on one end of each of the wires. Using a piece of a stick-on label, label the wires *Cu*, *Fe/Cu*, and *Zn/Cu* just below the loops.

4. Polish each wire and metal strip with steel wool.

5. Using the laboratory balance, measure the mass of the *Fe/Cu* wire. Record it in your data table.

PART 2–PLATING

6. Attach one end of a wire with alligator clips to the loop on the *Fe/Cu* wire. Attach the alligator clip on the other end of the wire to the negative (−) terminal of the battery.

7. Attach one end of the other wire with alligator clips to the iron strip. Clip the other end of this wire to the positive (+) terminal of the battery, as shown in **Figure 2** below. Keep the iron strip away from the copper wire.

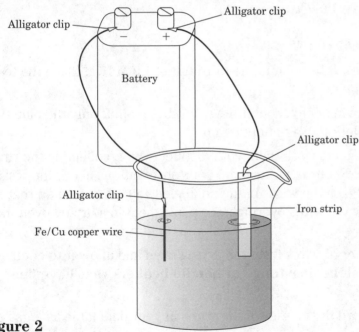

Figure 2

8. Pour 80 mL of the FeCl₃ solution into the *FeCl₃* beaker.

9. Using a stopwatch or clock with second hand to measure the time, immerse both the copper wire and the iron strip in the beaker, being careful to keep the alligator clips out of the solution. Use the loop to hang the wire on the beaker.

10. After about 5 min, remove the copper wire and the iron strip. Record the time elapsed to the nearest 1.0 s.

Electroplating for Corrosion Protection *continued*

11. Rinse the *Fe/Cu* wire with distilled water, collecting the rinse water in the *Waste* beaker. Record your observations about the *Fe/Cu* wire in your data table.

12. Hold the unplated *Zn/Cu* wire close to the plated *Fe/Cu* wire. Use a wax pencil to make a mark on the *Zn/Cu* and *Cu* wires at about the same level as the edge of the plating on the *Fe/Cu* wire. Measure and record the masses of these wires.

13. Repeat steps **6–11** with the zinc metal strip, zinc sulfate solution in the *ZnSO₄* beaker, and the wire labeled *Zn/Cu*. Be certain to plate the wire for exactly the same amount of time. Also be certain that the *Zn/Cu* wire is immersed up to the wax pencil mark.

14. Place the plated wires in the unlabeled 150 mL beaker so that they are not touching each other. Either allow the wires to dry in the beaker overnight, or place the beaker in a drying oven for 10 min. **Remember to use beaker tongs to handle all glassware that has been in the drying oven.**

15. After the wires have cooled, measure and record the masses of the *Fe/Cu* wire and the *Zn/Cu* wire in the data table.

16. In your data table rewrite the original mass of the *Cu* wire in the *New mass of wire (g)* row under the *Cu* column.

PART 3–TESTING REACTIVITY

17. Fill each of the test tubes about one-third full of 1.0 M HCl. Place the test tubes in a test-tube rack.

18. Place one of the wires into each of the test tubes so that only the plated parts are in the HCl solution. Wait about 5 min.

19. Remove the wires, and rinse them with distilled water, collecting the rinse in the *Waste* beaker. Record the time and your observations about the wires in your data table. Place the wires in the unlabeled 150 mL beaker so that they do not touch. Dry the wires overnight, or place them in a drying oven for 10 min.

20. Remove the beaker of wires from the drying oven and allow it to cool. **Remember to use beaker tongs to handle beakers that have been in the drying oven.**

21. Measure and record the masses of the wires in your data table as *Mass of wire after HCl (g)*.

22. Your teacher will provide separate disposal containers for each solution, each metal, and contents of the *Waste* beaker. The HCl from the test tubes can be poured into the same container as the contents of the *Waste* beaker.

TABLE 1: ELECTROPLATING DATA

	Cu	Fe/Cu	Zn/Cu
Mass of wire (g)	0.77	0.77	0.77
Plating time (s)	n/a	301.00	299.00
New mass of wire (g)	0.77	0.82	0.89
HCl time (s)	300.00	300.00	300.00
Mass of wire after HCl (g)	0.77	0.77	0.77

Observations

Students' data should also include observations about the wire. After

plating, the metals should be visible on the wire. The iron will look dark and

silvery, and the zinc will look silvery, but lighter than the iron. After

treatment with HCl, little of the plated metals will remain.

Analysis

1. **Applying Models** Write the equation for the half-reactions occurring on the metal strip and the copper wire in the $FeCl_3$ and $ZnSO_4$ beakers. Which is the anode in each one? Which is the cathode in each one?

iron strip: $Fe(s) \longrightarrow Fe^{3+}(aq) + 3e^-$, anode

copper wire: $Fe^{3+}(aq) + 3e^- \longrightarrow Fe(s)$, cathode

zinc strip: $Zn(s) \longrightarrow Zn^{2+}(aq) + 2e^-$, anode

copper wire: $Zn^{2+}(aq) + 2e^- \longrightarrow Zn(s)$, cathode

2. **Analyzing Data** How many grams and moles of iron and zinc were plated onto the *Fe/Cu* and *Zn/Cu* wires?
0.05 g Fe, 0.12 g Zn

9×10^{-4} mol Fe, 1.8×10^{-3} mol Zn

Conclusions

3. **Analyzing Results** Which metal or metal combination was the least reactive in HCl? Explain the basis for your conclusion.

The untreated copper wire was least reactive. When the zinc- and iron-plated

wires were placed in the HCl, bubbling occurred, indicating a reaction.

Electroplating for Corrosion Protection *continued*

4. **Evaluating Conclusions** What disadvantages relating to the use of the least reactive metal for the tanks can you think of?

Students' suggestions about the disadvantages of copper will vary. Possible

answers include the following: copper is soft and malleable, so the tanks

might be easily harmed; copper costs more than iron or zinc.

Extensions

1. **Research and Communication** Find out what measures are taken to try to prevent metal bridges and buildings from corroding, and prepare a chart to show the different methods, their relative costs, and their general uses.

Students will find a variety of methods in use. The Teflon used in the Statue

of Liberty prevents the iron and copper in the statue from coming into direct

contact. In this way, the redox reactions that led to the crumbling and corro-

sion of the previous iron framework can be greatly reduced. Other coatings

and paints work the same way to keep the reactants separated. In other

approaches, a sacrificial anode, which is made of a more active metal than

the main part of the item, is used. This approach is used on some ships'

hulls. Another possibility is to apply a small amount of voltage to an object

so that reduction rather than oxidation is favored.

Electric Charge

Teacher Notes

TIME REQUIRED 30 min

SKILLS ACQUIRED
Collecting data
Experimenting
Inferring
Identifying patterns
Predicting
Communicating

RATING Easy ←——1———2———3———4——→ Hard
Teacher Prep–1
Student Set-Up–1
Concept Level–2
Clean Up–1

THE SCIENTIFIC METHOD

Make Observations In the Procedure, the students observe the effect of the charged rod on the stream of water.

Form a Hypothesis In Conclusion questions 1 to 3, students make predictions and explain the reasoning for their predictions.

Analyze the Results In Analysis questions 1 to 3, students analyze how differences in charge or size of the water stream affect the results.

Communicate the Results When answering questions in the Procedure and Analysis questions 1 to 3 students communicate results by providing written answers to the questions.

MATERIALS

Materials for this lab activity can be purchased from Sargent-Welch. See the Lab Materials QuickList software on the **One-Stop Planner** CD-ROM for ordering instructions.

SAFETY CAUTIONS

Safety goggles and a lab apron must be worn at all times.

In case water splashes, use paper towels to mop up the water. At the sink, wring them out until they are only damp, and put them in the trash.

TIPS AND TRICKS

Students should be aware that high humidity makes charging the rod more difficult.

Discuss with students how electric charge is accumulated and subsequently discharged. Students should associate electron loss with positive charge accumulation and electron gain with negative charge accumulation.

Name _____ Class _____ Date _____

Electric Charge

When fur and a hard rubber rod are rubbed together, electrons are transferred between them. Both the fur and the rod become charged. When the charged rod comes close to a water stream, the rod induces a complementary charge on the side of the water stream closest to the rod, causing the water to move toward the rod. In this activity, you will attract a stream of water to a charged hard rubber rod.

OBJECTIVES

Observe and describe the effects of static charge on a stream of water.

Infer explanations for observed behaviors of the water stream.

MATERIALS

• hard rubber rod and fur pad, or acetate strip and wool swath

• faucet with cold running water

• 15 cm plastic ruler

 **Always wear safety goggles and a lab apron to protect your eyes and clothing.** If you get a chemical in your eyes, immediately flush the chemical out at the eyewash station while calling to your teacher. Know the locations of the emergency lab shower and eyewash station and the procedures for using them.

Procedure

1. Turn on a cold-water faucet. Adjust the stream to a diameter of about 1.5 mm. Make sure the stream is a steady flow.

2. Rub the rubber rod with the fur pad. Be sure that neither gets wet.

3. Holding the rod about 10 cm below and to the side of the faucet, slowly bring the rod within 3 cm of the water stream. Note your observations.

   The stream of water should bend toward the rod.

   _____

   _____

   _____

4. Move the rod closer to the water. Note your observations.

The bending of the stream should increase.

_____

_____

5. Repeat steps **2** through **4**. Note any changes in your initial observations.

Answers will vary, but students should observe a more strongly bent stream

because of the increased static charge.

_____

6. Adjust the faucet to produce a larger stream of water. Repeat steps **2** through **4**. Note your observations.

The water stream does not bend.

_____

_____

DISPOSAL

Return equipment to its proper place. Wash your hands thoroughly after all work is finished and before you leave the lab.

Analysis

1. Analyzing Results How does the distance between the rod and the water stream affect the results?

The smaller the distance between the rod and the stream, the more the

stream bends.

_____

2. **Applying Ideas**   How does a second charging of the rod affect the bending of the water stream? Give a reason for the change in behavior of the water stream.

Presuming the rod is still charged when the fur is stroked the second time,

the charge on the rod will be greater and the water stream will therefore

bend more the second time than the first time. A larger static charge induces

a stronger complementary charge on the water stream.

3. **Analyzing Results**   When the diameter of the water stream is increased, does the water bend more or less? Explain your observation.

Less; the charge induced by the rod must be spread over more water

molecules, so the effect on any one molecule is considerably less than in the

case of the thin stream.

Conclusions

1. **Applying Ideas and Predicting Outcomes**   Water is an electrically neutral polar molecule. Consider the structures of isopropyl alcohol and carbon tetrachloride. If the stream of liquid were isopropyl alcohol or carbon tetrachloride instead of water, would the results of the experiment be the same as the results obtained with water? Explain your reasoning.

No. Isopropyl alcohol molecules are electrically neutral and only slightly

polar, so the static charge on the rod would bend the alcohol stream ever so

slightly but not as much as water. Carbon tetrachloride molecules are

electrically neutral and nonpolar. The static charge on the rod would not

be strong enough to induce a stream of carbon tetrachloride to bend.

Electric Charge *continued*

2. Inferring Results and Designing Experiments

a. In this lab, you do not know the charge on the rubber rod. In the figures below, sketch several water molecules in the orientation that would explain the results you observed in this lab.

negative rod   positive rod   water stream

b. According to the orientations of the water molecules you predicted in (a), compare the direction that the water stream will bend using a negative rod with the way it will bend using a positive rod.

The stream will bend in the same direction (toward the rod) whether the

rod is positively charged or negatively charged.

_____

c. Design an experiment to test your hypothesis.

Answers will vary. Student responses should include a source and method

for producing a known positively charged object and a known negatively

charged object. Students should also indicate that the charged object will

be brought close to a water stream and the direction the stream is bent will

be observed.

3. Predicting Outcomes   When clothes are dried in an electric dryer, a dryer sheet is sometimes added to reduce static cling and to give the clothes a fresh smell. Suppose the rod in this experiment is rubbed with a dryer sheet before it is rubbed with the fur. Predict the effect on the water stream when the rod is brought near it. Explain your reasoning.

The water stream should not bend, because the chemical from the dryer

sheet has coated the rod and prevented a sufficient transfer of electrons

between the fur and rod.

_____

Electric Charge *continued*

4. Applying Concepts   During dry winter seasons, static cling is prevalent, affecting clothes and hair. Applying hair spray to a comb before pulling it through the hair can reduce or prevent static cling. Explain.

Answers will vary, but students should recognize that the coated surface

does not transfer electrons as readily as the noncoated surface, so the

concentration of charge is not sufficient to produce cling.

_____

_____

_____

_____

_____

A Cloth of Many Colors

Teacher Notes

TIME REQUIRED One 45-minute lab period

SKILLS ACQUIRED
Collecting data
Communicating
Experimenting
Identifying patterns
Inferring
Interpreting
Organizing and analyzing data

RATING
Easy ←——1——2——3——4——→ Hard

Teacher Prep–3
Student Set-Up–3
Concept Level–3
Clean Up–2

THE SCIENTIFIC METHOD

Make Observations Students collect data on direct and mordant dyeing.

Analyze the Results Analysis questions 1 and 2

Draw Conclusions Conclusions question 1 and Analysis question 2

Communicate the Results Analysis questions 1 and 2 and Conclusions question 1

MATERIALS

To prepare 1 L of 0.5 M $KAl(SO_4)_2 \cdot 12H_2O$ solution, observe the required safety precautions. Add 237 g of $KAl(SO_4)_2 \cdot 12H_2O$ to 200 mL distilled water in a volumetric flask. Gently agitate this mixture until all solute is dissolved. Add distilled water to exactly 1 L.

Dyeing solutions: To prepare a 0.2% solution of each dye, observe the required safety precautions. Add 2 g of the dye to 800 mL distilled water. Gently agitate this mixture until all dye is dissolved. Dilute to 1 L.

To prepare 0.25 M Na_2CO_3 solution, observe the required safety precautions. Add 31.0 g of $Na_2CO_3 \cdot H_2O$ to 200 mL distilled water in a volumetric flask. Gently agitate this mixture until all solute is dissolved. Add distilled water to exactly 1 L.

To prepare 0.5 M NaOH solution, observe the required safety precautions. Add 20.0 g of NaOH to 400 mL distilled water in a volumetric flask. Avoid dusting conditions. Gently agitate this mixture until all the solute is dissolved. Add distilled water to exactly 1 L.

Use 100% white cotton fabric as a source for sample material. Cut the fabric into 5 in. × 5 in. pieces, six pieces per group.

Dyes are available through scientific supply houses.

A Cloth of Many Colors *continued*

SAFETY CAUTIONS

Wear safety goggles, a face shield, impermeable gloves, and an apron when you prepare the alum, dye, Na_2CO_3, and NaOH solutions. Avoid dusting conditions. Work in a chemical fume hood known to be in operating condition and have another person stand by to call for help in case of an emergency. Be sure you are within a 30 s walk from a safety shower and eyewash station known to be in good operating condition.

In case of a spill, use a dampened cloth or paper towels to mop up the spill. Then rinse the cloth in running water in the sink and wring it out until it is only damp, and put it in the trash.

Caution students to exercise great care when handling heated materials.

Remind students that boiling water can scald. Emphasize that heated liquids need time to cool before removing fabric samples. Always use utility tongs to handle fabrics heated to boiling.

Remind students of the following safety precautions:

- Always wear safety goggles, gloves, and a lab apron to protect your eyes and clothing. If you get a chemical in your eyes, immediately flush the chemical out at the eyewash station while calling to your teacher. Know the location of the emergency lab shower and the eyewash stations and procedures for using them.

- Do not touch any chemicals. If you get a chemical on your skin or clothing, wash the chemical off at the sink while calling to your teacher. Make sure you carefully read the labels and follow the precautions on all containers of chemicals that you use. If there are no precautions stated on the label, ask your teacher what precautions you should follow. Do not taste any chemicals or items used in the laboratory. Never return leftover chemicals to their original containers; take only small amounts to avoid wasting supplies.

- Call your teacher in the event of a spill. Spills should be cleaned up promptly, according to your teacher's directions.

- Never put broken glass in a regular waste container. Broken glass should be disposed of properly.

DISPOSAL

Solutions may be flushed down the drain, using copious amounts of water.

TECHNIQUES TO DEMONSTRATE

Show students how to cut 1-in. square fabric pieces for distribution.

TIPS AND TRICKS

Plan additional time for students to wash fabric pieces and analyze fade test results (Part 2). Light fade tests should be planned for a minimum of one week; a longer time period yields more definitive results.

Assign each student group one dye solution. A dye may be assigned more than once.

Plan to have a clothesline for students to clothespin their samples for drying. Make a dishpan available in a sink area so that students can wash dyed fabrics.

Observe fabric samples in as strong a light as possible. You may choose to have students use fine-point forceps to remove threads from each of the fabric samples, prepare wet mounts, and make observations under low power (40×) of a compound microscope.

You may wish to point out that Allura red (Red 40) and Sunset yellow (Yellow No. 6) are food additives in Kool Aid® drink mixes.

A 10× magnifying glass is suggested for **step 20.** If possible, prepare wet mount standards as viewing samples for the class.

Observation of the individual fibers under a compound microscope should show that mordanted fibers appear coated, while direct-stained fibers appear infiltrated with dye.

Lightfastness depends on the amount of exposure time.

Monoazo control swatches should exhibit the most intensity of color when compared with colorfast or lightfast experimental groups. Methyl orange, alizarin, and Bismark Brown Y produce the most intense color. Students should observe only small differences between alizarin and Bismark Brown Y for dye intensity and relative fastness when compared against the respective controls.

A much greater difference is observed between these two dyestuffs and methyl orange. Sunset Yellow and Allura Red dyes do not bond well to cotton and do not exhibit fastness.

Name ______________________ Class ______________ Date ____________

Inquiry

A Cloth of Many Colors

A dye is a colored substance used to impart more or less permanent color to other substances. Dyes color most manufactured products, but their most important use is with textile fibers and fabrics. A dye must have an affinity for the substance it colors; it must enter the interior of the fiber and adhere. Starch and various other materials are applied to fabrics to increase their stiffness. These materials (called sizing) clog the pores of fibers and must be removed before the dye can reach the fibers. When a successful commercial dye resists fading when washed repeatedly, the dye is *colorfast.* When the dye also resists fading when exposed to direct light, it is *lightfast.* Not all dyes are suitable for use with every type of fabric. The chemistry of the fiber is critical to the binding of any dye: wool (polypeptide), cotton (cellulose), nylon (polyamide), dacron (polyester). A dye that is suitable for nylon may not dye cotton at all.

Cotton's fiber chemistry is versatile, making it an ideal medium for dye compounds and dyeing processes. Direct dyeing is the simplest process: a dye that is in solution attaches itself to the fiber of a fabric by direct chemical interaction. The majority of synthetic direct dyes are *azo* dyes, compounds that have at least one N=N linkage in the molecule. Diazo dyes have two N=N linkages and are excellent cotton fabric dyes. Interestingly, monoazo dyes, compounds with only one N=N linkage, do not bond well to cotton fiber.

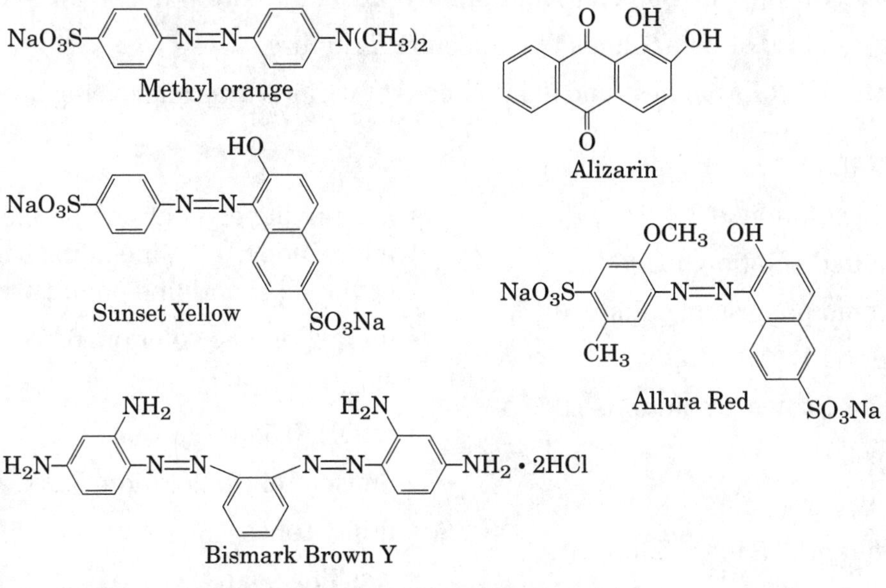

Figure 1: Structures of dyes used in this experiment

▌A Cloth of Many Colors *continued*

Another dye process is mordanting. The fabric is first treated with a heavy metal salt, the mordant, which binds with cotton fibers. When the dye is applied, it links with the mordant: sort of a "chemical handshake" among three individuals! Typical mordants are alum (potassium aluminum sulfate), copper(II) sulfate, or potassium dichromate. **Figure 2** shows how direct and mordant dyes bind to cellulose fibers in cotton fabrics.

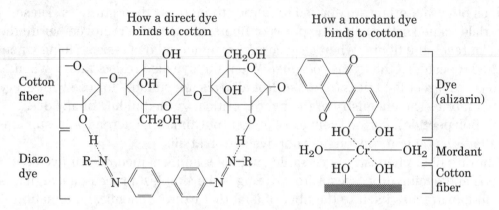

Figure 2: How direct and mordant dyes bond to cotton fibers

OBJECTIVES

Observe how direct dyeing and mordant dyeing color cotton fibers.

Identify a class of compounds by molecular structure.

Evaluate the colorfastness and lightfastness of various dyes and dyeing processes.

MATERIALS

- alizarin solution, 0.2%
- Allura Red solution, 0.2%
- aluminum potassium sulfate, 0.5 M
- beakers, 1 L (3)
- Bismark Brown Y solution, 0.2%
- clothesline
- clothespins
- cotton fabric, 5 in. × 5 in. (6)
- dishpan
- hot plate
- indelible ink laundry marker

- magnifying glass or compound microscope, 10×, including slides, cover slips, and fine-point tweezers
- methyl orange solution, 0.2%
- Na_2CO_3, 0.25 M
- NaOH, 0.5 M
- Sunset Yellow solution, 0.2%
- utility tongs
- washing soap
- white glue

⚠️ ⚠️ **Always wear safety goggles, gloves, and a lab apron to protect your eyes and clothing.** If you get a chemical in your eyes, immediately flush the chemical out at the eyewash station while calling to your teacher. Know the location of the emergency lab shower and eyewash station and the procedures for using them.

| A Cloth of Many Colors *continued*

☠ **Do not touch any chemicals.** If you get a chemical on your skin or clothing, wash the chemical off at the sink while calling to your teacher. Make sure you carefully read the labels and follow the precautions on all containers of chemicals that you use. If there are no precautions stated on the label, ask your teacher what precautions to follow. Do not taste any chemicals or items used in the laboratory. Never return leftovers to their original container; take only small amounts to avoid wasting supplies.

🜄 **Call your teacher in the event of a spill.** Spills should be cleaned up promptly, according to your teacher's directions.

Acids and bases are corrosive. If an acid or base spills onto your skin or clothing, wash the area immediately with running water. Call your teacher in the event of an acid spill. Acid or base spills should be cleaned up promptly.

◆ **Never put broken glass in a regular waste container.** Broken glass should be disposed of separately according to your teacher's instructions.

◆ **When using a hot plate, do not heat glassware that is broken, chipped, or cracked.** Use tongs or a hot mitt to handle heated glassware and other equipment because heated glassware does not always look hot.

Procedure
PART 1—REMOVING FABRIC SIZING

1. Put on safety goggles, gloves, and lab apron.

2. Using an indelible ink laundry marker, label each of the six fabric pieces with *one* of the following labels: Direct C, Mordant C, Direct L, Mordant L, Direct W, or Mordant W.

3. After labeling, place the pieces of cloth in a 1 L beaker. Add enough 0.25 M Na_2CO_3 to cover the cloths.

4. Place the beaker on a hot plate and bring the solution to a boil. Continue heating for 2 minutes. Turn off the heat and allow the solution to cool.

5. Using utility tongs, remove the treated cloths from the solution and rinse them thoroughly in running water. Wring dry. What other treatment could be employed to remove fabric sizing?

   machine washing with laundry detergent

PART 2—DIRECT DYEING

6. Select the fabric pieces from **step 5** that are labeled with the word *Direct*. Place them in a 1 L beaker.

7. Obtain from your teacher a liquid dye sample. Carefully pour 300 mL of the dye solution into the 1 L beaker. Place the beaker on a hot plate. Bring the solution to a boil and continue heating for 5 minutes. Allow the solution to cool.

A Cloth of Many Colors *continued*

8. Using utility tongs, remove the cloths from the dyeing solution. Rinse them under running water. Wring out the excess liquid. Using clothespins, hang the cloths up to dry. *Save the dye solution for use in* **step 14.**

PART 3—MORDANT DYEING

9. Select the fabric pieces from **step 5** that are labeled with the word *Mordant*. Place them in a second 1 L beaker. Pour 300 mL of 0.5 M $KAl(SO_4)_2 \cdot 12H_2O$ into this beaker.

10. Place the beaker on a hot plate and bring the solution to a boil. Continue heating for 5 minutes. Allow the solution to cool.

11. Using utility tongs, remove the cloth pieces, and rinse them under running water. Then place the cloth pieces into a third 1 L beaker. Pour 200 mL of 0.5 M NaOH into this beaker.

12. Place the beaker on a hot plate and bring the solution to a boil. Continue heating for 5 minutes. Allow the solution to cool.

13. Using utility tongs, remove the three cloth pieces. Rinse each under running water. Wring dry.

14. Place the cloth pieces from **step 13** into the beaker of dye solution you saved in **step 8.** Place the beaker on a hot plate, and bring the solution to a boil. Continue heating for 5 min. Allow the solution to cool.

15. Using utility tongs, remove each cloth piece from the solution. Rinse it under running water. Wring out the excess liquid. Using clothespins, hang each cloth up to dry.

PART 4—EVALUATING DYEING METHODS

16. Fill a dishpan 2/3 full of tap water. Dissolve a generous amount of soap in the water. Wash each of the two dried fabric pieces labeled with a "W" *at least* 10 times. Rinse thoroughly between washings. Redry.

17. Place each of the two dried fabric pieces labeled with an "L" on a window sill that has a direct southern exposure. Expose these pieces for at least a week.

18. *Do not* wash or place in direct sunlight the two dried fabric pieces labeled with a "C." What is the importance of these fabric pieces to your evaluations?

They serve as controls.

19. Cut the fabric pieces into 1-in. squares so that each lab group can paste a representative sample in **Table 1.**

A Cloth of Many Colors *continued*

TABLE 1 DYEING FABRICS

Glue fabric swatches in spaces indicated and compare.

DIRECT Colorfast	**DIRECT** Colorfast	**DIRECT** Colorfast	**DIRECT** Colorfast	**DIRECT** Colorfast
Glue fabric piece here	Glue fabric piece here	Glue fabric piece here	Glue fabric piece here	Glue fabric piece here
Methyl orange	Alizarin	Sunset Yellow	Allura Red	Bismark Brown
DIRECT Lightfast	**DIRECT** Lightfast	**DIRECT** Lightfast	**DIRECT** Lightfast	**DIRECT** Lightfast
Glue fabric piece here	Glue fabric piece here	Glue fabric piece here	Glue fabric piece here	Glue fabric piece here
Methyl orange	Alizarin	Sunset Yellow	Allura Red	Bismark Brown
DIRECT *Control*	**DIRECT** *Control*	**DIRECT** *Control*	**DIRECT** *Control*	**DIRECT** *Control*
Glue fabric piece here	Glue fabric piece here	Glue fabric piece here	Glue fabric piece here	Glue fabric piece here
Methyl orange	Alizarin	Sunset Yellow	Allura Red	Bismark Brown
MORDANT Colorfast	**MORDANT** Colorfast	**MORDANT** Colorfast	**MORDANT** Colorfast	**MORDANT** Colorfast
Glue fabric piece here	Glue fabric piece here	Glue fabric piece here	Glue fabric piece here	Glue fabric piece here
Methyl orange	Alizarin	Sunset Yellow	Allura Red	Bismark Brown
MORDANT Lightfast	**MORDANT** Lightfast	**MORDANT** Lightfast	**MORDANT** Lightfast	**MORDANT** Lightfast
Glue fabric piece here	Glue fabric piece here	Glue fabric piece here	Glue fabric piece here	Glue fabric piece here
Methyl orange	Alizarin	Sunset Yellow	Allura Red	Bismark Brown
MORDANT *Control*	**MORDANT** *Control*	**MORDANT** *Control*	**MORDANT** *Control*	**MORDANT** *Control*
Glue fabric piece here	Glue fabric piece here	Glue fabric piece here	Glue fabric piece here	Glue fabric piece here
Methyl orange	Alizarin	Sunset Yellow	Allura Red	Bismark Brown

A Cloth of Many Colors *continued*

20. Use a magnifying glass, or compound microscope and slides, to view individual fibers with the representative fabric treatments. Summarize your analysis in **Table 2**.

TABLE 2 ANALYSIS REPORT

Dye name	Dye type: monoazo, diazo, mordant	Coats fibers (yes/no)	Colors fibers (yes/no)	Colorfast rating: excellent, good, fair	Lightfast rating: excellent, good, fair
Methyl orange	monoazo	no	yes	good/fair	good/fair
Alizarin	mordant	yes	no	excellent/good	excellent/good
Sunset Yellow	monoazo	no	yes	fair/poor	fair
Allura Red	monoazo	no	yes	fair/poor	fair
Bismark Brown Y	diazo	no	yes	excellent/good	excellent/good

21. Clean all apparatus and your lab station. Return equipment to its proper place. Dispose of chemicals and solutions in the containers designated by your teacher. Do not pour any chemicals down the drain or put them in the trash unless your teacher directs you to do so. Wash your hands thoroughly after all work is finished and before you leave the lab.

Analysis

1. **Analyzing Methods** Which dye process is most effective at producing colorfast and lightfast dyed fabrics? Give reasons for your answer.

Students should conclude that mordanting is most effective against color

fading. Mordanted fibers should retain the dye as well as direct-dyed fibers.

Monoazo dyes (applied as a direct dyeing process) are not direct dyes and

will not exhibit the same dyeing properties.

Name _____ Class _____ Date _____

A Cloth of Many Colors *continued*

2. Analyzing Data View dyeing results. Which dye(s) appear to be the weakest? Which appear to be the strongest?

Answers will vary, but Bismark Brown Y should be the most intense. Allura

Red and Sunset Yellow are least intense.

Conclusions

1. Predicting Outcomes If you were the research chemist in a company that planned to market home-use dye products, how would you proceed?

A product design that incorporates a direct dyestuff, capable of dyeing

batches that can be heated over a stove, would most likely be commercially

successful based on both cost and results. Mordanting involves more

chemicals, is time consuming, and yields results that are not demonstrably

superior to that of direct dyeing.

The Slime Challenge

Teacher Notes
TIME REQUIRED 45 min

SKILLS ACQUIRED
Collecting data
Communicating
Experimenting
Identifying patterns
Inferring
Interpreting
Organizing and analyzing data

RATING
Easy ←——1——2——3——4——→ Hard

Teacher Prep–4
Student Set-Up–3
Concept Level–2
Clean Up–2

THE SCIENTIFIC METHOD

Make Observations Students observe the properties of several polymers.

Analyze the Results Analysis questions 1 to 5

Draw Conclusions Analysis questions 1 to 5

Communicate the Results Analysis questions 1 to 3 and 5

MATERIALS

To prepare 1 L of 5% polyvinyl alcohol (PVA) solution, heat approximately 700 mL of deionized water to 60–70°C. To avoid dusting, pre-wet the PVA (50 g) in about 200 mL of cold water, and mix to a slurry using a spatula. Slowly add *hot* water, using vigorous magnetic stirring. Once dissolution occurs, add deionized water to a final volume of 1 L. Cover the beaker with clear plastic wrap, and continue heating at 60–70°C until the solution clears; this may take as long as six hours. Do not exceed 80°C. The time for dissolution will vary depending on the molecular mass of the material; higher molecular masses have higher viscosities. See Tips and Tricks for additional information.

To prepare 1 L of white glue solution, mix 500 mL of white glue with 300 mL of deionized water, using vigorous magnetic stirring. Add additional deionized water to bring the final volume to 1 L.

To prepare 1 L of guar gum solution, slowly sprinkle 17 g of guar gum powder into 800 mL of *warm* water, using vigorous magnetic stirring. Add the powder slowly to avoid dusting and clumping. Once dissolution occurs, add deionized water to a final volume of 1 L.

To prepare 500 mL of 4% sodium borate solution, add 8 g of sodium borate decahydrate to 300 mL of deionized water, and stir until total dissolution occurs. Add additional deionized water to dilute the solution to 500 mL.

For the food coloring, use commercially available colors; green is the most popular slime color.

In procedure step **11,** the students will use a water-soluble marker to write on a piece of paper, then use the polymers to lift an imprint of the writing off the paper. This works best with colorless polymers.

SAFETY CAUTIONS

Read all safety precautions, and discuss them with your students.

• Safety goggles and an apron must be worn at all times.

• Although these polymers are made from nontoxic materials, caution students NOT to ingest any of these materials and to wash their hands after handling them.

• It is not advisable to allow students to remove the polymers from the lab; they may appear elsewhere in the school building.

TIPS AND TRICKS

An additional area of investigation may be to add iron oxide or iron filings—the amounts varying with experimenter's preference—to the slime samples to create Gak,™ another marketed toy polymer. Students can experiment with how Gak responds to various magnetic fields.

Have several lab groups prepare warm water baths needed for Procedure step **9.** For each warm water bath, fill an aluminum baking pan with water about 4 cm deep. Place the pan on a hot plate set on low heat. Place a thermometer in the water. Bring the water to about 60°C and maintain it at that temperature.

Discuss polymerization reactions and what is meant by cross-linking. Students should understand the concept of elasticity.

In step **7,** you may wish to have students press their thumbs into each sample and observe what happens. PVA and PVAc deform (stretch), then return to their original shape; guar gum does not deform, so students' fingers go through the material.

Polyvinyl alcohol comes in a variety of molecular masses. The medium-viscosity PVA (120,000–150,000 g/mol) is best for formulating 3–5% polymer solutions because it does not have to be heated. Higher viscosity (molecular mass) product requires heating. In all cases, pre-wetting the material, using cold water to form a slurry, is recommended to avoid lumps.

White glue contains polyvinyl acetate (PVAc).

DISPOSAL

Set out three plastic dishpans, labeled *1, 2,* and *3*. The numbered slime solutions should go in the corresponding numbered dishpans. Put the contents of each container in a large plastic bag, label it with the contents, and put it in the trash.

Name _____ Class _____ Date _____

The Slime Challenge

Slime® is a trade name for a toy material first marketed commercially in the 1970s. Its success dramatizes just how entertaining polymer chemistry can be. Polymers are chains of macromolecules formed by the union of five or more identical combining units (monomers). A cross-linking agent can be added to enhance a polymer's characteristics. In 1839, Charles Goodyear noticed that if natural rubber and sulfur were heated together they became "linked," allowing the rubber to remain firm and elastic at high temperatures instead of soft and sticky. It is the same for slime.

In this laboratory activity, you will evaluate how the cross-linking reaction of sodium borate affects certain physical characteristics of the three polymers described in the **Information Table.** Sodium borate decahydrate, $Na_2B_4O_7 \cdot 10H_2O$, is the salt of a strong base (NaOH) and a weak acid $B(OH_3)$. In water, the weak acid hydrolyzes to form a borate ion, which reacts with a hydroxyl group. Polyvinyl alcohol has hydroxyl groups attached to the main polymer chain, so it can link to the borate ion. Following linkage, most of the lattice structure is filled with water molecules, and the carbon-to-oxygen-to-boron bonds are easily broken and re-formed. **Figure A** shows the cross-linking reaction between polyvinyl alcohol and sodium borate.

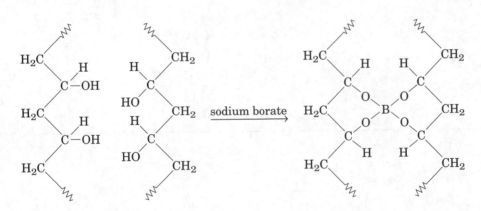

Figure A

The Slime Challenge *continued*

Information Table

Polymer name	Natural or synthetic	Repeating monomeric unit	Polymer type
Guar gum	natural		polysaccharide

galactose (35%)
mannose (65%)

Polymer name	Natural or synthetic	Repeating monomeric unit	Polymer type
Polyvinyl alcohol (PVA)	synthetic		thermoplastic elastomer

Polymer name	Natural or synthetic	Repeating monomeric unit	Polymer type
Polyvinyl acetate (PVAc)(white glue)	synthetic		thermoplastic elastomer

Polysaccharide—a combination of nine or more monosaccharides (simple sugars, like $C_6H_{12}O_6$) linked together.

Thermoplastic elastomer—a high-chain polymer that softens when exposed to heat and has the ability to be stretched to twice its original length and to retract rapidly when released.

OBJECTIVES

Prepare slime by using sodium borate as a cross-linking agent with various natural and synthetic polymers.

Compare the physical properties of prepared slime variants.

Evaluate which polymer makes the best slime.

MATERIALS

- 4% sodium tetraborate (sodium borate) solution, 10 mL
- 5% polyvinyl alcohol (PVA) solution, 30 mL
- 50% white glue solution, 30 mL
- 1.7% guar gum solution, 30 mL
- 50 mL graduated cylinders, 3
- aluminum baking pans
- food coloring, assorted colors
- water-soluble marker
- ice cubes
- 4 oz plastic cups, 3
- 30 cm plastic ruler
- spatula
- thermometer, nonmercury
- wax pencil
- hot plate
- $\frac{1}{4}$ in. diameter wooden dowel, 8 in.

  Always wear safety goggles and a lab apron to protect your eyes and clothing. If you get a chemical in your eyes, immediately flush the chemical out at the eyewash station while calling to your teacher. Know the locations of the emergency lab shower and the eyewash station and the procedures for using them.

Do not touch any chemicals. If you get a chemical on your skin or clothing, wash the chemical off at the sink while calling to your teacher. Make sure you carefully read the labels and follow the precautions on all containers of chemicals that you use. If there are no precautions stated on the label, ask your teacher what precautions you should follow. Do not taste any chemicals or items used in the laboratory. Never return leftovers to their original containers; take only small amounts to avoid wasting supplies.

Call your teacher in the event of a spill. Spills should be cleaned up promptly, according to your teacher's directions.

Never put broken glass in a regular waste container. Broken glass should be disposed of properly.

Never stir with a thermometer, because the glass around the bulb is fragile and might break.

When using a Bunsen burner, confine long hair and loose clothing. If your clothing catches on fire, WALK to the emergency lab shower and use it to put out the fire. Do not heat glassware that is broken, chipped, or cracked. Use tongs or a hot mitt to handle heated glassware and other equipment because hot glassware does not look hot.

Procedure

PART 1: MAKING SLIME AND ITS VARIANTS

1. Use the wax pencil to label cups *1*, *2*, and *3*. Using separate graduated cylinders, pour 30 mL of guar gum solution into cup *1*, 30 mL of polyvinyl alcohol into cup *2*, and 30 mL of white glue into cup *3*.

2. (Optional) Add one drop of food coloring to each polymer solution.

3. To the guar gum in cup *1*, rapidly add 5.0 mL of sodium borate solution, and stir the mixture briskly for 2 min. Repeat with the monomers in cups *2* and *3*.

PART 2: EVALUATING THE PHYSICAL PROPERTIES OF SLIME VARIANTS

4. **Observe the appearance:** Poke each slime sample with your index finger. Record your observations below.

guar gum

When poked, the material is easily penetrated; the slime is colorless (unless

food coloring has been added) and looks like goo.

polyvinyl alcohol (PVA)

When poked, the material is not easily penetrated and the finger may bounce

back; the slime is colorless (unless food coloring has been added) and looks

like a blob.

polyvinyl acetate (PVAc)

When poked, the material is not easily penetrated and is tougher than PVC

polymer, and the finger may bounce back; the slime is white (unless food

coloring has been added) and looks like a blob.

5. For the polyvinyl alcohol and polyvinyl acetate slime samples, form each polymer into a ball and place the balls about 10 cm apart on your lab table. Observe for 5 min. Record your observations below.

polyvinyl alcohol

It flows slowly and becomes a flat blob.

polyvinyl acetate

It flows slowly and becomes a flat blob.

6. For the polyvinyl alcohol and polyvinyl acetate slime samples, form each polymer into a ball, and drop it from a height of 30 cm. Roughly measure how high each bounces. Record your measurements below.

polyvinyl alcohol __15 cm__

polyvinyl acetate __12 cm__

7. Elastic recoil: Take the slime sample from cup *1* in both hands. *Slowly* stretch it to twice its original length, then release one end. Does the sample return rapidly to its original length? Record your observations below. Repeat for the slime samples in cups *2* and *3*.

guar gum

The slime is not rigid enough to pull apart; it just falls as you attempt to

stretch it.

polyvinyl alcohol

The slime snaps back.

polyvinyl acetate

The slime snaps back.

8. Take the slime sample from cup *1* in both hands. *Quickly* stretch it. Record your observations below. Repeat for the slime samples in cups *2* and *3*.

guar gum

The slime is not rigid enough to pull apart; it just falls as you attempt to

stretch it.

polyvinyl alcohol

The slime tears; each side of the tear is smooth.

polyvinyl acetate

The slime tears; each side of the tear is smooth.

9. **Comparing thermal elasticity:** Pair with another group. One group can do procedure step **9** and the other can do step **10**. Both groups observe the testing and make their own observations. Place the cups containing the polyvinyl alcohol and polyvinyl acetate slime samples into an aluminum baking pan containing *warm* water (approximately 60°C). Let the samples warm in the water for 10 min.

For the polyvinyl alcohol slime sample, quickly form the polymer into a ball, and drop it from a height of 30 cm. Roughly measure how high it bounces. Repeat this procedure for polyvinyl acetate. Record your measurements below.

polyvinyl alcohol __12 cm__

polyvinyl acetate __9 cm__

10. Place the cups containing the polyvinyl alcohol and polyvinyl acetate slime samples into an aluminum baking pan containing *ice* water. Let the samples cool in the water for 10 min.

For the polyvinyl alcohol slime sample, quickly form the polymer into a ball, and drop it from a height of 30 cm. Roughly measure how high it bounces. Repeat this procedure for polyvinyl acetate. Record your measurements below.

polyvinyl alcohol __7 cm__

polyvinyl acetate __4 cm__

11. Polyvinyl alcohol and polyvinyl acetate slime samples will pick up ink from paper. Draw a picture or write your name backwards on paper, with a water-soluble marker. Press the ball of slime onto the paper for only a split second. The design on the paper will be transferred to the slime. (The slime will stick to the paper if left on too long.)

12. Flatten the polyvinyl alcohol and polyvinyl acetate slime samples. Place one on top of the other and observe for 5 min. Record your observations below.

The two slimes appear to bond to each other; they will not separate when an

attempt is made to pry them apart.

DISPOSAL

13. Clean all apparatus and your lab station. Return equipment to its proper place. Dispose of slimes, chemicals, and solutions in the containers designated by your teacher. Do not pour any chemicals down the drain or put them in the trash unless your teacher directs you to do so. Wash your hands thoroughly after all work is finished and before you leave the lab.

The Slime Challenge *continued*

Analysis

1. Analyzing Data From your observations, which polymer do you think made the best slime? Explain the reason(s) for your answer.

Answers may vary, but most students will say that the guar gum polymer

made the best slime because it was the gooiest looking.

2. Analyzing Data Which polymer sample(s) exhibited the property of elastic recoil?

PVA and PVAc should exhibit the property, with it more pronounced in PVA

than in PVAc. The natural polymer does not; it is not an elastomer.

3. Analyzing Data Do all cross-linked polymers rapidly return to their original length following stretching? Justify your answer, using the polymer characteristics you observed.

No, elastomers PVA and PVAc exhibit stretch elasticity; the nonelastomer

guar gum does not.

4. Analyzing Data Summarize the thermal elasticity of the polyvinyl alcohol and polyvinyl acetate slime samples after they were heated and cooled, based on a comparison of the heights the polymer balls bounced.

Warm temperatures reduced the elasticity of both polymers; neither bounced

as high after warming. Cooler temperatures also reduced the elasticity of the

polymers. The polyvinyl alcohol sample may crumble when the students try

to roll it into a ball.

5. Analyzing Data What do you think caused the phenomenon observed in Procedure step **12**?

The two polymers appeared to have formed cross-links with one another

since the polymers appeared to bond to each other.

All Fats Are Not Equal

Teacher Notes

TIME REQUIRED One 45-minute lab period

SKILLS ACQUIRED
Collecting data
Communicating
Experimenting
Identifying patterns
Inferring
Interpreting
Organizing and analyzing data

RATING
Easy ◄——— 1   2   3   4 ———► Hard

Teacher Prep–3
Student Set-Up–3
Concept Level–2
Clean Up–3

THE SCIENTIFIC METHOD

Make Observations Using common household fats, students will determine the degree of saturation in fatty acids and relate the degree of saturation to melting point.

Analyze the Results Analysis questions 1 to 4

Draw Conclusions Analysis questions 1 and 2

Communicate the Results Analysis questions 1 and 2

MATERIALS

Wear safety goggles, disposable polyethylene gloves, and an apron when you prepare the iodine tincture solution. Work in a chemical fume hood known to be in operating condition and have another person stand by to call for help in case of an emergency. Be sure you are within a 30 second walk from a working safety shower and eyewash station.

To prepare 100 mL of tincture of iodine, add 7 g of I_2 crystals and 5 g of KI to 5 mL of distilled water, and dilute to 100 mL with denatured ethyl alcohol.

SAFETY CAUTIONS

Read all safety precautions, and discuss them with your students.

Safety goggles, gloves, and an apron must be worn at all times.

Broken glass should be disposed of in a clearly labeled box lined with a plastic trash bag. When the box is full, close it, seal it with packaging tape, and set it next to the trash can for disposal.

CAUTION: Tincture of iodine is a flammable liquid and a powerful stain. Avoid open flames or sparks. Have students handle with care.

Remind students of the following safety precautions:

- Always wear safety goggles, gloves, and a lab apron to protect your eyes and clothing. If you get a chemical in your eyes, immediately flush the chemical out at the eyewash station while calling to your teacher. Know the location of the emergency lab shower and the eyewash stations and procedures for using them.

- Do not touch any chemicals. If you get a chemical on your skin or clothing, wash the chemical off at the sink while calling to your teacher. Make sure you carefully read the labels and follow the precautions on all containers of chemicals that you use. If there are no precautions stated on the label, ask your teacher what precautions you should follow. Do not taste any chemicals or items used in the laboratory. Never return leftover chemicals to their original containers; take only small amounts to avoid wasting supplies.

- Call your teacher in the event of a spill. Spills should be cleaned up promptly, according to your teacher's directions.

- Never put broken glass in a regular waste container. Broken glass should be disposed of properly.

DISPOSAL

Dispose of solid foodstuffs as an inert solid waste. Collect liquid oil mixtures and mix thoroughly with 25 to 50 mL of liquid detergent to disperse the oils, and pour down the drain with copious amounts of water.

TIPS AND TRICKS

A single graduated cylinder can be used if it is washed and rinsed between samples to avoid mixing of oils.

If a change does not occur in the sunflower or cod liver oil (these oils have the highest iodine number and will shift color first) within 10 minutes at room temperature, have students place all the samples in a water bath and apply low heat. Within a few minutes, both these oils should turn colorless.

It is important to add initially the same amount of iodine to each of the oils. If the addition of iodine does not turn all oils immediately red-violet, continue adding iodine drop by drop.

Depending on the brand used, some students may rank sunflower oil as having the lowest ranking in **Table 1.**

In Part 2, it is important to start heating at a temperature below 32°C so that an accurate melting point can be determined for all solids.

Discuss the iodine number and how it indicates unsaturation. The iodine number is the number of grams of iodine that react with 100 grams of fat. Iodine numbers for the oils in Part 1 are: cod liver oil 135–165, sunflower oil 125–135, corn oil 110–130, peanut oil 90–100, coconut oil 6–10.

Student's data are qualitative, not quantitative. Although students do not calculate an actual iodine number, the color shift observed and the number of drops of iodine solution used allow an assessment of a generalized ranking. Results may vary according to product brands.

Students may benefit from a discussion of the table in the introduction.

Review the water-bath setup.

All Fats Are Not Equal

In a saturated fatty acid, each carbon atom is connected to its neighbors by single bonds, while in an unsaturated fatty acid, some carbon atoms are connected by double bonds. The number of carbon-carbon double bonds in a molecule is the substance's degree of unsaturation. The degree of unsaturation and the total number of carbon atoms in the fatty acid chains determine the differences between fats and oils. For example, myristic acid is a solid at room temperature according to the **Information Table,** but oleic acid, which has one carbon-carbon double bond, is a liquid. Similarly, you should notice an increase in melting points as you move from myristic acid to stearic acid because the number of carbon atoms increases. In general, at room temperature, fats are solids and oils are liquids. Therefore, you might predict a fat to be mostly saturated fatty acids and an oil to be mainly unsaturated fatty acids.

To determine the degree of unsaturation, scientists test for the amount of iodine that reacts with a 100 g sample of fat or oil. This value is the iodine number. The higher the value of the iodine number, the greater the amount of unsaturation in the fat or oil. When I_2 is added to the colorless fat or oil, the mixture appears red violet, like I_2. During the reaction, the color of the mixture fades as I_2 adds to the carbon-carbon double bond, producing a colorless product.

INFORMATION TABLE—REPRESENTATIVE FATTY ACIDS OF DIETARY FATS AND OILS

Fatty acid	Melting point (°C)	Class (saturated or unsaturated)	Molecular structure
Myristic acid	58	Saturated	CH_3—$(CH_2)_{12}$—CO_2H
Palmitic acid	63	Saturated	CH_3—$(CH_2)_{14}$—CO_2H
Stearic acid	71	Saturated	CH_3—$(CH_2)_{16}$—CO_2H
Oleic acid	16	Monounsaturated	CH_3—$(CH_2)_7$—$CH{=}CH$—$(CH_2)_7$—CO_2H
Linoleic acid	−5	Polyunsaturated	CH_3—$(CH_2)_4$—$CH{=}CH$—CH_2—$CH{=}CH$—$(CH_2)_7$—CO_2H
Linolenic acid	−11	Polyunsaturated	CH_3—CH_2—$CH{=}CH$—CH_2—$CH{=}CH$—CH_2—$CH{=}CH$—$(CH_2)_7$—CO_2H

OBJECTIVES

Determine the degree of unsaturation in fatty acids.

Relate how melting point indicates the degree of saturation.

All Fats Are Not Equal *continued*

MATERIALS

- beaker, 500 mL
- beaker tongs
- beakers, 25 mL (6)
- butter
- coconut oil
- cod liver oil
- corn oil
- gloves
- graduated cylinders, 25 mL (5)
- hot plate
- lab apron
- milk chocolate, 1 in. × 0.25 in. piece
- peanut oil
- ring stand
- safety goggles
- soft margarine
- spatula
- stick margarine
- sunflower oil
- tablespoon
- test-tube rack
- test tubes, medium (10)
- thermometer
- thermometer clamp
- tincture of iodine
- vegetable shortening
- wax pencil

  Always wear safety goggles, gloves, and a lab apron to protect your eyes and clothing. If you get a chemical in your eyes, immediately flush the chemical out at the eyewash station while calling to your teacher. Know the location of the emergency lab shower and eyewash station and the procedures for using them.

Do not touch any chemicals. If you get a chemical on your skin or clothing, wash the chemical off at the sink while calling to your teacher. Make sure you carefully read the labels and follow the precautions on all containers of chemicals that you use. If there are no precautions stated on the label, ask your teacher what precautions to follow. Do not taste any chemicals or items used in the laboratory. Never return leftover chemicals to their original containers; take only small amounts to avoid wasting supplies.

Call your teacher in the event of a spill. Spills should be cleaned up promptly, according to your teacher's directions.

Never put broken glass in a regular waste container. Broken glass should be disposed of separately according to your teacher's instructions. **Never stir with a thermometer because the glass around the bulb is fragile and might break.**

All Fats Are Not Equal *continued*

Procedure

PART 1—DETERMINING THE DEGREE OF UNSATURATION IN COMMERCIALLY AVAILABLE OILS

1. Put on safety goggles, gloves, and a lab apron.

2. Use a wax pencil to label five individual test tubes "Peanut oil," "Sunflower oil," "Corn oil," "Cod liver oil," and "Coconut oil."

3. Using a graduated cylinder, measure 10 mL of peanut oil and pour it into the appropriately labeled test tube. Set the test tube in a test-tube rack. Do the same for each of the other four oils.

4. Add two drops of tincture of iodine to each labeled test tube. *Carefully* swirl each test tube to disperse the iodine into small droplets. Return the test tube to the test-tube rack.

5. Let each mixture of oil and iodine stand for at least 10 minutes. Note the time it takes for any color change to occur *after* adding the iodine. Record both the time and color change in **Table 1.**

6. Determine an "unsaturation ranking" for this set of oil samples based on whether a color change occurs (red violet to colorless). If a color change occurs, record the elapsed time. Record your ranking in **Table 1.**

PART 2—DETERMINING THE MELTING POINT OF FOODSTUFFS AND THE DEGREE OF FATTY ACID SATURATION

7. Use a wax pencil to label six individual beakers "Vegetable shortening," "Butter," "Corn oil," "Margarine," "Soft margarine," and "Chocolate."

8. Measure a *level* tablespoon (5 g) of each soft food sample, and place it in its correspondingly labeled beaker. Use a spatula to help level each soft food sample. Place the piece of chocolate in its beaker.

9. Using **Figure 1** as a guide, prepare a water bath. Place one of the beakers prepared in **step 8** in the water bath. Heat on the hot plate's low setting, so that the temperature of the water gradually increases from room temperature. Monitor the temperature. Record the temperature at which the food sample liquefies completely in **Table 2.** Using beaker tongs, remove the warmed sample from the water bath. Repeat for each food sample. Record the room temperature for corn oil.

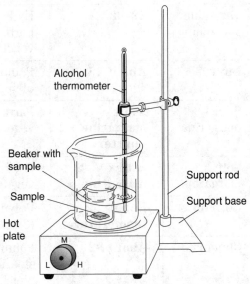

Figure 1

| All Fats Are Not Equal *continued*

10. For each sample tested, review the fatty acid ingredients listed in **Table 2** and your melting-point data. Then rank each food sample from highest saturated fatty acid content to lowest saturated fatty acid content. Record this ranking in **Table 2**.

11. Clean all apparatus and your lab station. Return equipment to its proper place. Dispose of your materials according to your teacher's directions. Dispose of chemicals and waste oils in containers designated by your teacher. Do not pour any chemicals or oils down the drain or put them in the trash unless your teacher directs you to do so. Wash your hands thoroughly after all work is finished and before you leave the lab.

TABLE 1 DETERMINING THE DEGREE OF UNSATURATION IN OILS

Oil type	Number of iodine (I_2) drops	Time to change color (min)	Color change (√)	Analysis ranking (most unsaturated to least) 1–5
Peanut oil	2		√	4
Sunflower oil	2		√	2
Corn oil	2		√	3
Cod liver oil	2	**Fastest**	√	1
Coconut oil	2	**No change**	√	5

TABLE 2 MELTING POINT AND DEGREE OF SATURATION

Food sample	Melting point (°C)	Fatty acid ingredient(s)	Analysis ranking (most unsaturated to least)
Vegetable shortening	22–32	Hydrogenated and partially hydrogenated vegetable oils	5
Butter	**About 15**	Palmitic acid (29%), oleic acid (27%)	1
Stick margarine	**>soft margarine <butter**	Partially hydrogenated vegetable oils	2
Soft margarine	**<stick margarine**	Partially hydrogenated vegetable oils	3
Corn oil	**Room temperature**	Polyunsaturated acids (34%), oleic acid (50%)	6
Chocolate	**About 32**	Palmitic acid (24%), stearic acid (35%), oleic acid (38%)	4

Answers will vary. Margarines should have lower melting points than butter.

All Fats Are Not Equal *continued*

Analysis

1. **Analyzing Data** Examine your entries in **Table 1.** What trend do you observe in vegetable oils regarding unsaturated fatty acid side chains?

   <u>Most vegetable oils are high in unsaturated fatty acids.</u>

   _____

2. **Analyzing Data and Applying Concepts** Coconut oil is a major ingredient in many nondairy creamers and other prepared foods. If an individual is trying to reduce saturated fat intake, would a nondairy creamer containing coconut oil be a good choice? Explain.

   <u>No. Cream (butterfat) and coconut oil are both high in saturated fatty acids.</u>

   _____